A GUIDE TO CLASSIFICATION IN GEOLOGY

ELLIS HORWOOD SERIES IN GEOLOGY
Editor: D.T.DONOVAN, Professor of Geology, University College London

This series aims to build up a library of books on geology which will include student texts and also more advanced works of interest to professional geologists and to industry. The series will include translations of important books recently published in Europe, and also books specially commissioned.

FAULT AND FOLD TECTONICS
W. JAROSZEWSKI, Department of Geology, University of Warsaw
A GUIDE TO CLASSIFICATION IN GEOLOGY
J. W. MURRAY, Professor of Geology, University of Exeter
THE CENOZOIC ERA
C. POMEROL, Professor, University of Paris VI.
Translated by D. W. HUMPHRIES, Department of Geology, University of Sheffield, and E. E. HUMPHRIES. Edited by Professor D. CURRY and D. T. DONOVAN, University College London
INTRODUCTION TO PALAEOBIOLOGY: GENERAL PALAEONTOLOGY
B. ZIEGLER, Professor of Geology and Palaeontology, University of Stuttgart, and Director of the State Museum for Natural Science, Stuttgart

BRITISH MICROPALAEONTOLOGICAL SOCIETY SERIES

This series, published for the British Micropalaeontological Society, will gather together knowledge for a particular faunal group for specialist and non-specialist geologists alike. The scope of the series has been broadened to include the common elements of the fauna, whether index or long-ranging species, and to convey a broad impression of the fauna and allow the reader to identify common species as well as those of restricted stratigraphical range.

The synthesis of knowledge presented in the series will reveal its strengths and prove its usefulness to the practicing micropalaeontologist, and those teaching and learning the subject. By identifying some of the gaps in the knowledge, the series will, it is believed, promote and stimulate further active research and investigation.

A STRATIGRAPHICAL INDEX OF CALCAREOUS NANNOFOSSILS
Editors: G. B. HAMILTON, Consultant Micropalaeontologist, and A. R. LORD, Department of Geology, Unversity College London
A RESEARCH MANUAL OF FOSSIL AND RECENT OSTRACODS
Editors: R. HOLMES BATE, British Museum of Natural History, London, E. ROBINSON, Department of Geology, University College London, and L. SHEPPARD, British Museum of Natural History, London
STRATIGRAPHICAL ATLAS OF FOSSIL FORAMINIFERA
Editors: J. W. MURRAY, Professor of Geology, University of Exeter, and G. JENKINS, The Open University
MICROFOSSILS FROM RECENT AND FOSSIL SHELF SEAS
Editors; J. W. NEALE, Professor of Micropalaeontology, University of Hull, and M. D. BRASIER, Lecturer in Geology, University of Hull

A GUIDE TO CLASSIFICATION IN GEOLOGY

J. W. MURRAY, D.I.C., B.Sc., Ph.D., F.G.S.
Professor of Geology
University of Exeter

QE
425.4
M87
1980

51061

ELLIS HORWOOD LIMITED
Publishers · Chichester

Halsted Press: a division of
JOHN WILEY & SONS
New York · Chichester · Brisbane · Toronto

First published in 1981 by
ELLIS HORWOOD LIMITED
Market Cross House, Cooper Street, Chichester,
West Sussex, PO 19 1EB, England

The publisher's colophon is reproduced from James Gillison's drawing of the ancient Market Cross, Chichester.

DISTRIBUTORS:

Australia, New Zealand, South-east Asia:
Jacaranda-Wiley Ltd., Jacaranda Press,
JOHN WILEY & SONS INC.,
G.P.O. Box 859, Brisbane, Queensland 40001, Australia

Canada:
JOHN WILEY & SONS CANADA LIMITED
22 Worcester Road, Rexdale, Ontario, Canada

Europe, Africa:
JOHN WILEY & SONS LIMITED
Baffins Lane, Chichester, West Sussex, England

North and South America and the rest of the world:
Halsted Press: a division of JOHN WILEY & SONS
605 Third Avenue, New York, N.Y. 10016, U.S.A.

British Library Cataloguing in Publication Data
Murray, J. W.
   A guide to classification in geology.
   1. Geology – Classification
   I. Title
   550'.1'2   QE7   80–41094
ISBN 0-85312-193-1 (Ellis Horwood Ltd., Publishers – Library Edition)
ISBN 0-85312-319-5 (Ellis Horwood Ltd., Publishers – Student Edition)
ISBN 0-470-27090-X (Halsted Press)

Typeset in Press Roman by Ellis Horwood Ltd.
Printed in Great Britain by Butler & Tanner Ltd., Frome, Somerset.

COPYRIGHT NOTICE
© 1981 J. W. Murray/Ellis Horwood Ltd.
All Rights Reserved. No part of this publication may be reproduced, stored in a retrieval system, or transmitted, in any form or by any means, electronic, mechanical, photocopying, recording or otherwise, without the permission of Ellis Horwood Limited, Market Cross House, Cooper Street, Chichester, West Sussex, England.

Table of Contents

Author's Preface 9

Chapter 1 – Introduction 11

Chapter 2 – Sediments and Sedimentary Rocks
  2.1 Major groups of sedimentary rocks 13
  2.2 Terrigenous sediments and rocks........... 14
      2.2.1 Rudites – conglomerates and breccias... 14
      2.2.2 Sandstones – arenites and wackes...... 19
      2.2.3 Siltstones........................ 22
      2.2.4 Lutites – mudstone, mudrock, claystone. 22
  2.3 Volcanic sedimentary deposits 25
      2.3.1 Pyroclastic rocks................... 28
      2.3.2 Hydroclastic and Hyaloclastic rocks 31
      2.3.3 Phreatomagmatic rocks.............. 31
      2.3.4 Autoclastic rocks................... 31
      2.3.5 Alloclastic rocks 32
      2.3.6 Terrigenous sediments rich in volcanic fragments 32
  2.4 Residual deposits....................... 33
  2.5 Organic deposits 33
      2.5.1 Carbonaceous deposits 33
      2.5.2 Carbonate sediments and carbonate rocks (limestones)...................... 35
      2.5.3 Dolomitic limestones and dolomites 44
      2.5.4 Siliceous sediments and rocks 45

Table of Contents

   2.6   Chemical precipitates 46
       2.6.1   Evaporites......................... 46
       2.6.2   Phosphorites 46
       2.6.3   Ironstones........................ 46
       2.6.4   Caliche and calcrete................ 46
       2.6.5   Silcrete 47
       2.6.6   Chert and flint.................... 47
   2.7   Special groups......................... 47
       2.7.1   Deep sea sediments 47
       2.7.2   Process name rock types 48
   2.8   References............................ 49

Chapter 3 – Igneous Rocks
   3.1   Plutonic rocks......................... 55
       3.1.1 $M < 90\%$........................ 55
       3.1.2 $M > 90\%$ – ultramafic rocks......... 56
       3.1.3   Charnockites..................... 65
   3.2   Hypabyssal rocks...................... 65
   3.3   Volcanic rocks........................ 68
       3.3.1   Glassy rocks 72
       3.3.2   Igneous rock series 72
       3.3.3   Metamorphosed volcanic rocks 73
   3.4   References............................ 74

Chapter 4 – Metamorphic Rocks
   4.1   Names based on texture 77
   4.2   Names based on composition 78
   4.3   Gneiss............................... 80
   4.4.   References............................ 83

Chapter 5 – Mixed Rocks and Rock Associations
   5.1   Migmatite 84
       5.1.1   Structural types................... 85
       5.1.2   Metatexite and diatexite 86
   5.2   Ophiolite............................. 86

- 5.3 Rodingite 87
- 5.4 Kimberlite.......................... 87
- 5.5 References.......................... 87

Chapter 6 – Stratigraphic Classification
- 6.1 Lithostratigraphy 89
- 6.2 Biostratigraphy 90
- 6.3 Chronostratigraphy 90
- 6.4 References.......................... 93

Chapter 7 – Engineering Geology 94
- 7.1 Rock materials 97
- 7.2 Rock masses 100
- 7.3 Core logging 102
- 7.4 Aggregates......................... 102
- 7.5 References......................... 103

Index................................... 105

Author's Preface

Most professional geologists have specialist knowledge of only part of the subject. When they read papers outside their field of specialisation they may encounter problems with unfamiliar terminology. The problem is even greater for students and amateurs who are in the process of building up their knowledge. The majority of modern textbooks concentrate primarily on processes and mechanisms of producing rocks. My aim has therefore been to provide a precis of the classification systems in common use in geology in the hope that this small book will prove a useful source of reference.

I wish to thank Mr Ellis Horwood for encouraging me to write this book and Mrs G. F. Murray for typing the manuscript. Professor D. T. Donovan, Mr P. Grainger and Dr M. Stone kindly offered comment on various chapters.

Exeter
December 1979

Chapter 1

Introduction

To classify is defined as 'to arrange or distribute in classes according to a method or system' (*Shorter Oxford English Dictionary*). Accepting this definition, a classification does three basic things:
- it sets out criteria for distinguishing between the items being classified
- it allows grouping of similar items in an hierarchical scheme of classes
- it establishes a scheme of nomenclature.

In geology, classification plays an important role in the organisation of data. Rocks and structures need to be defined by standard schemes of nomenclature in order to assist communication between geologists. There have been many attempts to standardise nomenclature, but not all classification schemes are universally accepted. This proliferation of systems presents no problems so long as authors state clearly which classification scheme they are using. Moreover, it is doubtful whether it is desirable or necessary to employ a totally standard classification system, as new research inevitably reveals alternative ways of approaching the same problem.

Some geologists believe that the ideal classification is one that is purely descriptive. However, many geological classifications are genetic. For instance, rocks are classified, according to the mode of formation, into igneous, sedimentary, and metamorphic classes. In many cases individual rock names not only describe the rock but also define their chemical composition, texture and grain size, and the environment of formation, for example, quartz arenite, a sedimentary rock composed of detrital grains 0.0625 - 2 mm in diameter, more than 95% of which are quartz.

With so many variables involved it is not surprising that different authors place greatest emphasis on different variables, which in turn leads to disagreement. Thus, the term 'greywacke' fell into disrepute because it meant different things to different people: to some it was a muddy sandstone whereas to others the product of a turbidity current. However, not all muddy sandstones are the product of turbidity currents, and not all turbidity currents deposit muddy sand. The problem can be solved by using 'greywacke' for a muddy sandstone and, for those laid down by turbidity currents, by using the term 'turbiditic greywacke.'

Some classifications, although founded on sound logic, have not been accepted, often because they involve cumbersome terminology. The ideal classification is simple to use, provides a workable system of nomenclature, and is acceptable to those using it. Chapters 2 - 7 represent a compilation of those classifications proposed by international working parties (for example, igneous rocks) and others commonly used. No attempt has been made to include all available classification schemes.

Chapter 2

Sediments and sedimentary rocks

Sediment: an unconsolidated deposit of either detritus derived from pre-existing rock or new minerals formed by chemical or biochemical processes or organically formed materials.

Sedimentary rock: a lithified sediment, including those sediments which have undergone total mineralogical reorganisation under conditions of low temperature and low pressure (but excluding true metamorphism).

2.1 MAJOR GROUPS OF SEDIMENTARY ROCKS

A commonly used subdivision is as follows:

Terrigenous deposits - composed of detrital transported material.

Volcanic sedimentary deposits - composed of fragmented volcanic material extruded penecontemporaneously with sediment deposition. Sometimes this group is termed 'pyroclastic' but this type is only one of several volcanic sedimentary deposits.

Residual deposits - left after chemical weathering of rock, for example laterite.

Organic deposits - composed of the remains of plants or animals (all limestones are conveniently included here even though some contain only a low organic component).

Chemical precipitates - notably evaporite mineral deposits whether primary or secondary (diagenetic) in origin.

An additional category has been added here to cater for process name rock types and deep sea sediments.

2.2 TERRIGENOUS SEDIMENTS AND ROCKS

Terrigenous sediments consists of the detritus of pre-existing rocks which has been produced by weathering (physical and chemical), has been transported by wind, water or ice, has suffered varying degrees of abrasion, and is finally deposited.

The processes of lithification include compaction, cementation, and the chemical and mineralogical alteration of unstable minerals, for example feldspars being altered to clays.

The classification of terrigenous sediments and rocks is based on their composition and grain size.

The principal components are fragments (clasts) of rock, quartz and feldspar, together with clay minerals.

Grain size varies from boulder to clay (Figs. 2.1 and 2.2)

2.2.1 Rudites - conglomerates and breccias

These are made up of clasts >2 mm in diameter.

Gravels and conglomerates

In the majority of cases the clasts are derived from a variety of source rocks separate from the site of deposition (extraformational). However, in some instances penecontemporaneous reworking of a sediment during deposition produces cohesive fragments which are redeposited (intraformational).

∅	diameter mm	Particle		Size Terms (rock group name)		Common types of terrigenous sediment
-8	256	Boulder	Gravel	Rudaceous (rudite)	Psephitic (psephite)	Conglomerate or breccia
-6	64	Cobble	Gravel	Rudaceous (rudite)	Psephitic (psephite)	Conglomerate or breccia
-2	4	Pebble	Gravel	Rudaceous (rudite)	Psephitic (psephite)	Conglomerate or breccia
-1	2	Granule	Gravel	Rudaceous (rudite)	Psephitic (psephite)	Conglomerate or breccia
0	1	Very coarse sand		Arenaceous (arenite)	Psammitic (psammite)	Sandstone
1	0.5	Coarse sand		Arenaceous (arenite)	Psammitic (psammite)	Sandstone
2	0.25	Medium sand		Arenaceous (arenite)	Psammitic (psammite)	Sandstone
3	0.125	Fine sand		Arenaceous (arenite)	Psammitic (psammite)	Sandstone
4	0.0625	Very fine sand		Arenaceous (arenite)	Psammitic (psammite)	Sandstone
4	0.0625	Silt		Argillaceous or Lutaceous (lutite)	Pelitic (pelite)	Siltstone
8	0.0039	Clay		Argillaceous or Lutaceous (lutite)	Pelitic (pelite)	Mudstone

Fig. 2.1 — Sedimentary size grades and terms. The ∅ (phi) scale is a logarithmic transformation of the Wentworth scale: $\emptyset = -\log_2 d$, where d is the diameter in mm (Krumbein, 1934, 1936). Particle data after Wentworth (1922).

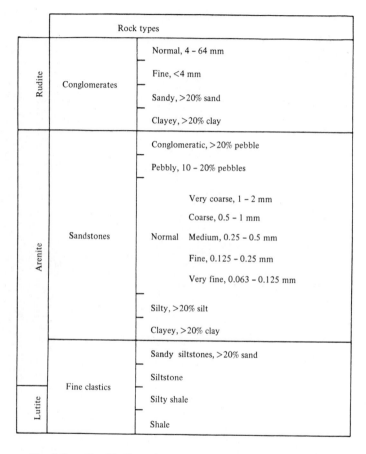

Fig. 2.2 – Classification of terrigenous rocks based on grade size (after Krynine, 1948).

Terrigenous Sediments and Rocks

		Clasts			
		Rounded		Angular	
Size grade	mm	Unlithified	Lithified	Unlithified	Lithified
Boulder	256	Gravel	Conglomerate	Rubble or scree	Breccia or sharpstone conglomerate
Cobble	64				
Pebble	4				
Granule	2				

Fig. 2.3 – Classification of rudites based on particle size and shape.

There are various approaches to the classification of these sediments. Some are descriptive, for example cobble conglomerate; some are genetic, for example fanglomerate, tillite. When the origin of a deposit is in doubt it is wise to use purely descriptive terms, for example, 'petromict conglomerate' rather than 'fanglomerate.' Schermerhorn (1966) introduced the term *mixtite* for those diamictites which are not clearly recognisable as either tillites or tilloids. For the diagnostic features of tills and tillites see Harland *et al.* (1966).

Rubble, scree, breccia and sharpstone conglomerate.
The term *breccia* is applied to any rock composed of angular fragments. Sedimentary breccias have been termed *sharpstone conglomerates* by Pettijohn (1975) to distinguish them from breccias of volcanic and tectonic origin.

Support	Matrix	Clasts	Mechanism of deposition	Rock name	Rock group
clast	<15%	>90% quartz or chert (mature)	Water	oligomict or orthoquartzitic conglomerate	Orthoconglomerates
clast	<15%	>90% metastable, limestone, granite, etc. (immature)	Water	petromict or polymict conglomerate fanglomerate-alluvial fan	Orthoconglomerates
matrix	>15%, laminated	angular, poorly sorted, varied rock types	clasts (dropstones) dropped into sediment from floating ice, plants, etc.	conglomeratic mudstone or argillite	Paraconglomerates or diamictites (unlithified: diamicton)
matrix	>15%, not laminated	poorly sorted; if large rock masses, olistoliths	glacial (ice)	tillite (unlithified: till or 'boulder clay')	Paraconglomerates or diamictites (unlithified: diamicton)
matrix	>15%, not laminated	poorly sorted; if large rock masses, olistoliths	non-glacial, usually mass flow especially subaqueous	tilloid (if mappable: olistostrome)	Paraconglomerates or diamictites (unlithified: diamicton)

Fig. 2.4 – Classification of extraformational conglomerates (based partly on Pettijohn, 1975).

2.2.2 Sandstone: arenites and wackes

The term *sandstone* refers to a rock composed of sand sized grains (625 μm - 2 mm in diameter). Strictly speaking there is no connotation of composition, and the grains may be of terrigenous or volcanic origin or of autochthonous carbonate. Nevertheless, sandstone is commonly taken to mean a rock composed of terrigenous quartz grains. If there is a noticeable proportion of feldspar grains, then the term *feldspathic sandstone* may be used, implying a mainly quartz sand with some feldspar.

The grain size within the spectrum of the sand grade may be specified by the use of the prefixes fine, medium or coarse.

Some sandstones are said to be clean because they lack a muddy matrix, but others have a small to large proportion of mud. Such rocks may be loosely called *muddy sandstones* or *greywackes*. The latter term has had a long and controversial history (Cummins, 1962; Dott, 1964). Clean quartz sandstones are regarded as compositionally mature because they are composed only of mineralogically stable grains. Muddy sandstones and greywackes are said to be immature because they contain mineralogically unstable grains (feldspar) or the products of their destruction (clay minerals).

Textural maturity of sandstones increases from low in muddy, poorly sorted, angular sands to high in well sorted, well rounded sands.

Folk advocated using the prefixes immature, mature, etc., with the rock name, for example mature quartzite, but not in the case of arkoses and greywackes which are invariably immature and therefore need not have this specified.

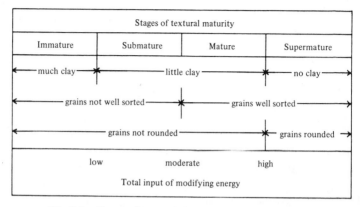

Fig. 2.5 — Stages of textural maturity (after Folk, 1951).

A simple classification of sandstones suitable for field use is shown in Fig 2.2 The commonly used petrographic classifications are those of Dott (1964) and Pettijohn, Potter and Siever's modification (1972). A basic subdivision is into arenites or clean sands (Fig. 2.6) and wackes or dirty sands in which the matrix may be primary or be due to secondary breakdown of unstable minerals (Fig 2.7).

Arkose is used for those feldspathic arenites with more feldspar grains than lithic fragments.

Lithic arenites may be described by the principal lithic type, for example chert arenite.

The upper limit of matrix size is taken as 30 μm, but different authors use different limits.

Hybrid sandstones

Some sandstones contain, in addition to detrital clastic grains, other grains which are produced within the environment of deposition, that is, autochthonous grains.

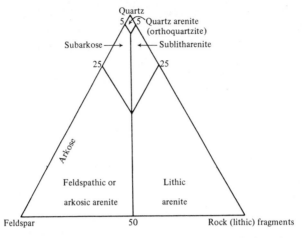

Fig. 2.6 — Subdivision of arenites (clean sandstones, <15% of matrix <30µm). Texturally mature or supermature. (based on Dott, 1964, and Pettijohn *et al.*, 1972).

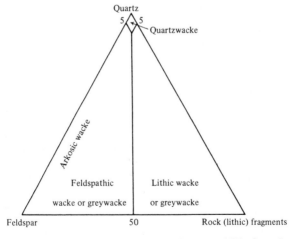

Fig. 2.7 – Subdivision of wackes (dirty sandstones, >15% of matrix <30µm). Texturally submature or immature. (based on Dott, 1964, and Pettijohn *et al.*, 1972).

Calcarenaceous sands: a mixture of detrital quartz and sand-sized carbonate of chemical or biochemical origin. Such sands grade into bioclastic and oolitic limestones. The use of the term 'calcarenaceous' is to distinguish these grain sands from those consisting of detrital grains with calcareous cement (Pettijohn, 1957).

Greensand or glauconitic sand: detrital sands with glauconite grains.

Mixed terrigenous and volcanic sandstones

Sandstones containing significant quantities of penecontemporaneous volcanic detritus are termed *tuffaceous sandstones* or *tuffites*.

2.2.3 Siltstones

Strictly speaking **silt** refers to grains having a diameter of 0.0039 – 0.0625 mm. In practice silt is often used as a descriptive term for a muddy fine sand.

Siltstone should be used for lithified detrital silt *sensu stricto*.

Loess is an unstratified, unconsolidated silt.

2.2.4 Lutites - mudstone, mudrock, claystone

These are sediments composed of clay minerals and detrital grains less than 0.0039 mm in diameter.

Kaolinite-rich clays are freshwater deposits.

Tonsteins form in association with plant remains.

Fireclays have refractory qualities (they will withstand temperatures of $1600°C$ without fusion). Fireclays and seatearths are commonly found beneath coal seams and are regarded as palaeosoils. Seatearths lack the refractory properties of fireclays and not all are of clay grade.

Dominant clay mineral					
Clay mineral mixture		Kaolinite		Montmorillonite	
Unlithified	Lithified	Unlithified	Lithified	Unlithified	Lithified
clay or mud	claystone or mudstone or mudrock or shale (if laminated or fissile)	kaolinitic clay or pipe clay or china clay or ball clay	kaolinitic claystone or tonstein	montmorillonite clay	bentonite or Fuller's earth
		fire clay or seatearth			

Fig. 2.8 – Classification of lutites.

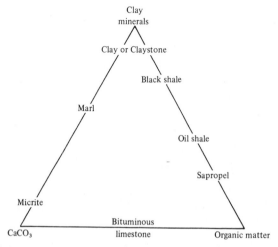

Fig. 2.9 – Triangular diagram to show the relationships between nomenclature and composition of lutites (after Selley, 1976).

Montmorillonite-rich clays and their equivalents represent decomposed volcanic ash.

See Lewan (1978) for a laboratory classification of clays.

Hybrid lutites

Fine grained sediments may contain admixtures of clay minerals, clay sized calcium carbonate and organic matter.

Marl – a mixture of clay and 40 - 60% clay sized calcium carbonate. When lithified, a *marlstone*. Some varieties suitable for cement making are termed *cementstones* or *hydraulic limestones*.

Black shale – otherwise known as *organic shale* or *carbonaceous shale*. It is rich in organic matter and commonly in pyrite too.

Tuffaceous mudstone – mudstone containing a significant proportion of reworked penecontemporaneous volcanic detritus.

2.3 VOLCANIC SEDIMENTARY DEPOSITS

The classification of sediments composed of debris of volcanic rocks is a matter of controversy (see Fisher, 1966), and a working group of the IUGS is at present looking into the problems. There is debate about the definition of basic terms, such as volcanic and pyroclastic, and there are differing viewpoints on the relative importance of the processes of fragmentation of the debris and its subsequent deposition. Part of the problem relates to the intermediate state of these deposits between igneous rocks proper and terrigenous sediments proper.

During the eruption of a volcano lava may be extruded, gas (including water vapour) is vented, and associated with its release fragments of volcanic rocks are ejected into the air (or water if a subaqueous eruption).

The lavas are igneous rocks and are considered in Chapter 3.

The gas is dispersed and leaves no direct record.

The ejected material, collectively known as *tephra,* will be deposited around the volcano as a subaerial or subaqueous deposit, commonly said to be pyroclastic. However, in addition, other processes operate which lead to the formation of deposits composed of fragments of volcanic rocks.

Use of the terms volcanic and volcaniclastic

'Volcanic' is an adjective derived from the noun volcano; examples of use are, volcanic rock, that is, rock produced by a volcano, and volcanic cone, that is, a cone produced by volcanic processes. The term is sometimes informally treated as a noun, for example Permian volcanics, meaning Permian volcanic rocks.

'Volcaniclastic' was coined by Fisher (1961) to group together all rocks composed of fragments of volcanic rocks, regardless of whether they are the products of volcanism alone or reworked volcanic material. Wright and Bowes (1963) have criticised the use of this word, pointing out that it is an adjective and means breakage by volcanic processes. For this reason, it is wrong to apply the word to sediments composed of reworked volcanic material. The latter are sedimentary conglomerates, breccias or lithic arenites depending on grain size.

Classifications and names for rocks and sediments produced directly or indirectly by volcanic activity have been set up to do several things:
- (a) to give a descriptive name to the rock type,
- (b) to identify the processes by which fragmentation of the clasts took place,
- (c) to identify the mechanism and environment of deposition,
- (d) to identify post-depositional diagenetic/metamorphic changes.

No one system of nomenclature can achieve all these. In fact even in modern volcanic environments not all these aims can be achieved, for example most of our knowledge of submarine eruptions is inferred from the deposits rather than based on first-hand observation of the processes operating (for obvious reasons).

While it may be useful to have a system of nomenclature which defines not only the rock type but also how and where it formed, it is not useful if misinterpretations of the processes of formation lead to misidentification. Clearly what is needed is a system of nomenclature which is descriptive and applicable to all rocks regardless of their method of origin (See Fig 2.10). For those rocks

mm	Non-genetic descriptive rock name	Method of fragmentation			
		PYROCLASTIC volcanic explosion	HYDROCLASTIC OR HYALOCLASTIC rapid cooling by water	PHREATOMAGMATIC water causes explosion
	x breccia or x conglomerate	pyroclastic breccia (angular clasts) or pyroclastic agglomerate (rounded clasts)	hydroclastite or hyaloclastite	
64 —		— — —		
32 —		lapillistone		
2 —		— — —	(Sometimes called 'palagonite tuff')	
	x Sandstone or x arenite or x wacke	tuff (unweathered)		hyalotuff
0.0625	x siltstone	bentonite		
0.0039	x mudstone	(weathered)		

mm	Method of fragmentation			
	AUTOCLASTIC flow of lava, etc.	ALLOCLASTIC other methods	Terrigenous sediment rich in volcanic debris
	autoclastic volcanic breccia varieties: friction-breccia flow-breccia explosion-breccia	alloclastic volcanic breccia varieties: intrusion-breccia tuffisite-breccia explosion-breccia intrusive-breccia	lahar-paraconglomerate (mudflow deposits of volcanic debris)
64 —			
32 —			
2 —			pépérite
		tuffisite	tuffaceous sandstone
0.0625			tuffaceous siltstone
0.0039			tuffaceous mudstone

Fig. 2.10 – Classification of rocks composed of fragments of volcanic rocks. Where the composition of the volcanic rock is known the appropriate prefix should be used in place of 'x' (column 2) and also, for example, in andesitic agglomerate, etc.

where the conditions of formation are clear a genetic system of nomenclature can be used in addition or alone.

References of general use are Fisher (1961, 1966), Honnorez and Kirst (1976), Peltz (1971), Pettijohn (1975), Sparks and Walker (1973), Wentworth and Williams (1932), Wright and Bowes (1963).

2.3.1 Pyroclastic rocks

These result from primary deposition of material ejected from a vent by a volcanic eruption. The ejected material is known collectively as *tephra*. Bombs are partly molten on ejection. *Achneliths* are small particles that are molten on ejection but solid by the time they descend to earth. Their size terms are shown in Fig 2.11.

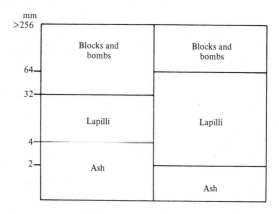

Fig. 2.11 — Grade size terms for tephra (left column after Wentworth and Williams, 1932; right column after Fisher, 1961).

Other terms used to describe pyroclastic debris are 'pumice' and 'scoria' — vesiculated juvenile magma; and 'shards' — fragments of pumice vesicle walls.

Sec. 2.3] Volcanic Sedimentary Deposits

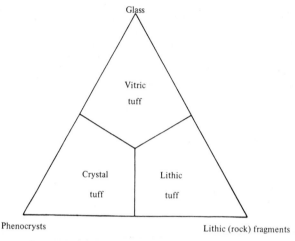

Fig. 2.12 — Triangular diagram of tuff varieties (after Pettijohn, Potter and Siever, 1972).

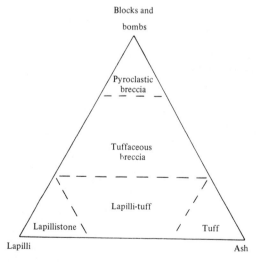

Fig. 2.13 — Triangular diagram of pyroclastic rocks. The boundaries between the rock types are subjective (after Fisher, 1966).

Varieties of tuff include:

Sintered tuff – following deposition a tuff may be fused (sintered) by heat from a later lava flow.

Welded tuff – during deposition internally supplied heat may cause fusion, and the welded tuff may flow like a lava. Welded tuffs are common in ignimbrites. Many 'rhyolites' are welded tuffs.

Bentonite – a chemically altered tuff (see Grim & Güven, 1978).

Mechanisms of emplacement of pyroclastic deposits

1. *Fall deposits.* These show an exponential decrease in grain size and deposit thickness with increasing distance from the vent. They mantle the topography, and may be internally stratified. Such deposits settle out of the air after explosive ejection from the vent. Where fall deposits accumulate under water they may or may not undergo sedimentary reworking.

2. *Flow deposits.* These accumulate mainly on flat ground or in depressions and form homogenous rock bodies. They lack internal stratification and the tephra are unsorted. Sparks and Walker (1973) use the term *pyroclastic flow* for the actively moving flow and *ignimbrite* for the resultant rock body, regardless of whether or not it is welded. Pyroclastic flows travel as dense fluidised masses of gas and tephra (mainly pumice). The aqueous analogue is a mudflow.

3. *Ground surge deposits.* These mantle the topography but their thickness is not uniform (cf. pyroclastic flow deposits). They have internal stratification which is inclined to the top and bottom of the deposit and often shows cross bedding. Unlike ignimbrites, ground surge

deposits are heterogenous, and are generally < 1 m thick (cf. >10 m for ignimbrites). Sparks and Walker (1973) believe the ground surge to be of low density and high velocity. The deposits show features comparable with those of river flood deposits. Because they are thin their chances of preservation are poor. Sparks and Walker (1973) suggest that a nuée ardente is made up of pyroclastic flow and a ground surge. The former follows valleys and the latter is less restricted so that although the two start from the same source, the deposits may be geographically separated.

2.3.2 Hydroclastic and Hyaloclastic rocks

Fragmentation is caused by the quenching effect of water or ice on hot glass. *Hyaloclastites* are developed particularly in association with pillow lavas.

Some authors use the term *palagonite tuff* for fine grained rocks of this type. Palagonite is a by-product of chemical reaction between hot lava and water.

2.3.3 Phreatomagmatic rocks

Water from the sea or a caldera lake may enter a magma chamber causing an explosion which ejects glassy fragments to form *hyalotuffs*.

2.3.4 Autoclastic rocks

Autoclastic volcanic breccia – 'forms by fragmentation of semi-solid and/or solid lava by explosive disruption of the gases contained within the lava or by the movement of the lava. It includes friction-breccia and some explosion-breccias' (Wright and Bowes, 1963).

Friction-breccia – 'forms by the disruption of lava by

further movement after part of the mass has congealed. It includes beccias formed within a volcanic plug and those formed at the surface by rising plugs or domes' (Wright and Bowes, 1963).

Flow-breccia – 'a type of friction breccia which forms by the flow of an unconfined lava'. (Wright and Bowes, 1963).

Autoclastic explosion-breccia – 'forms by the disruption of a semi-solid mass of igneous material by the explosion of cognate gases'. (Wright and Bowes, 1963).

2.3.5 Alloclastic rocks

Alloclastic volcanic breccia – 'forms by the fragmentation of any rock by any form of volcanic activity beneath the surface of the earth. It includes intrusion-breccia, explosion breccia, and intrusive breccia'. (Wright and Bowes, 1963).

Intrusion-breccia – 'forms by the forcible intrusion of magma into a country rock'. (Wright and Bowes, 1963).

Tuffisitic breccia – 'a type of intrusion-breccia in which the matrix is intrusive tuff (tuffisite) and in which the brecciation was produced by the intrusion of the tuffaceous material'. (Wright and Bowes, 1963).

Explosion breccia – 'forms by gas explosion in confined spaces beneath the surface'. (Wright and Bowes, 1963).

Tuffisite – intrusive tuff occurring as pipes or dykes, emplaced by fluidization.

2.3.6 Terrigenous sediments rich in volcanic fragments.

Fisher (1961, 1966) included these in his volcaniclastic group but many authors would prefer to treat them as primarily terrigenous. The volcanic component is normally indicated by the prefix 'tuffaceous'. Lahar deposits, that is, mudflows of volcanic debris, are paraconglomerates. Beach gravels composed of penecontemporaneous volcanic rocks are conglomerates.

Péperite, the product of intrusion of basaltic magma into wet sediments, has angular clasts of glass set in a sediment matrix. This unusual rock-type should perhaps be regarded primarily as a terrigenous sediment although the method of fragmentation is hyaloclastic.

2.4 RESIDUAL DEPOSITS

The chemical weathering of rocks on land leads to a breakdown of minerals. Some material is removed in solution. The remainder is left as a residual deposit. Soils are residual deposits but the only ones of particular interest to geologists are:

Terra Rosa – a ferruginous red soil representing the insoluble residue of underlying carbonate rocks.

Laterite – highly weathered residue rich in aluminium oxides and hydrated iron. Pisolithic and pipe structures are sometimes developed.

Bauxite – extremely weathered residue rich in aluminium oxide minerals.

2.5 ORGANIC DEPOSITS

2.5.1 Carbonaceous deposits

Modern sediments rich in organic materials are:

Peat – in situ accumulation of woody plant remains especially in freshwater swamps, (humus).

Sapropel – fine grained sediments rich in algal material. Fossil deposits rich in carbonaceous material can be classified in terms of three original components, clay, humus or peat, and sapropel (Fig 2.14).

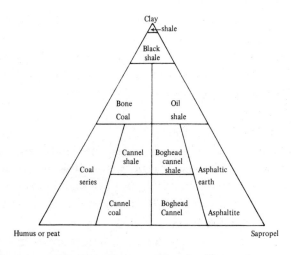

Fig. 2.14 — Classification of carbonaceous sediments according to their source components (after Pettijohn, 1975).

Black shale has a low proportion of carbon, is black in colour and commonly contains pyrite.

Bone coal contains a high proportion of clay and a low proportion of peaty material.

Oil shale is the equivalent rock with oil derived from sapropel.

The coal series is graded according to rank, that is, degree of alteration from original peat.

Cannel and boghead coals have medium to high proportions of sapropel in addition to detrital peat (Moore, 1968). Boghead coal is also known as torbanite.

Rank	Type	Brightness	Colour
low	peat	dull	brown
	soft brown coal		,,
	hard brown coal		,,
	bituminous hard coal	bright	black
high	anthracite	very bright	,,

Rank and characteristics of coal types (based on Teichmüller and Teichmüller, 1968).

2.5.2 Carbonate sediments and carbonate rocks (limestones)

No formal scheme of classification for all types of carbonate sediment has so far been proposed.

Many authors use the standard size grades sand, silt, mud, with the prefix carbonate or lime, for example carbonate sand, lime sand, etc. Alternatively where the sand-sized particles are of shell the term *skeletal* may be applied, for example skeletal sand. Other authors use terms such as 'foraminiferal mud,' 'grapestone sand,' etc.

Two classifications used in two tropical shallow water carbonate areas are given in Fig 2.15 and 2.16.

Lees (1975) proposed a scheme of names for skeletal sands and gravels based on composition but with environmental significance:

Foramol – Foraminifera, mollucs, barnacle, bryozoan and calcareous red algal debris. Temperate seas.

Chloralgal – calcareous green algae but no corals. Extreme salinity environments.

Chlorozoan – hermatypic corals and other groups. Tropical seas.

Because of profound diagenetic changes (Bathurst, 1971, p.102) it is difficult to relate the classification of

% fines	support	facies	environment
<25	grain	oolite (>80% oöids) oolitic (oöids and other grains) grapestone (grapestone > oöids)	shelf lagoon sands
		coralgal (coral and algal sand)	outer platform sands
>25	matrix	lime mud 1. skeletal mud 2. pellet mud	shelf lagoon muddy sands

Fig. 2.15 – Classification of Bahama Bank modern carbonate sediments (after Imbrie and Purdy, 1962).

sediment	major grain type
Sands 90%>63µm	ooidal bivalve compound grain/bivalve large perforate foraminiferal imperforate foraminiferal/pelletoidal gastropod coral/algal
muddy sands 10-50%>63µm	bivalve gastropod
mud 90%<63µm	bivalve imperforate foraminiferal/gastropod

Fig. 2.16 – Classification of Persian Gulf modern carbonate sediments (after Wagner and van der Togt, 1973).

soft carbonate sediments to either the Folk (Imbrie and Purdy, 1962) or Dunham (Wagner and van der Togt, 1973) classifications.

Deep sea carbonate sediments are considered on p.47.

Organic Deposits

Limestone is the group term given to lithified carbonate sediments.

In terms of significance it is of the same order as terrigenous rocks so use of limestone without specifying the type is very imprecise.

The principal schemes of classification used are those of Folk and Dunham.

Folk classification of limestones (Folk, 1959, 1962)

Limestones are made up of three components:

allochems – sand or gravel sized carbonate particles

micrite – microcrystalline ooze, the carbonate equivalent of clay

sparry calcite – normally a chemically deposited, pore-filling cement

Four types of allochems are volumetrically important. In each case of allochems the abbreviation used by Folk as part of the limestone name is given in parentheses.

Intraclast (intra): 'fragments of penecontemporaneous, usually weakly consolidated carbonate sediment that have been eroded from adjoining parts of the sea bottom and redeposited to form a new sediment'. (Folk, 1959, p.4). The emphasis is on reworking within the area of deposition. The term includes 'grapestones' but excludes fragments of older limestones eroded from land areas (calciliths).

Pellets (pel): 'rounded, spherical to elliptical or ovoid aggregates of microcrystalline calcite ooze, devoid of any internal structure' (Folk, 1959, p.6). Many of these are the faecal pellets of marine invertebrates. Some authors use the term peloid which includes pellets as well as other fine-grained carbonate grains (fro example shell fragments which have been altered to micrite by diagenetic processes).

Ooids or ooliths (oo): sand-sized spherical or near spherical grains having a nucleus and one or more concentric layers of carbonate crystallites. Modern marine ooids are normally aragonitic and the crystallites are arranged tangentically or perpendicular to the concentric layers. Fossil ooids are normally calcitic and have radially arranged crystallites.

Fossils (bio): calcareous hard parts of animals and plants.

Micrite (micr): Micrite is a contraction of microcrystalline calcite and refers to calcite grains in the size range 1-4μm. It is analogous to modern carbonate mud, i.e. aragonite mud. Micrite probably represents diagenetically modified aragonite mud.

Sparry calcite cement (spar): This forms a pore-filling cement. The crystal size shows considerable variation, from around 4μm and upwards. Spar is characteristically clear by contrast with the subtranslucent appearance of micrite.

Folk recognised three major limestone families (based on the proportions of allochems, micrite and sparry calcite cement), and two others (autochthonous reef rocks and replacement dolomites).

In comparison with terrigenous sediments, allochems are equivalent to sand or pebbles, micrite to clay matrix and sparry calcite cement to void (that is primary porosity). The relative proportions of micrite and sparry calcite cement reflect the energy input of the depositional environment and hence the textural maturity of the sediment. The importance of the allochems is in the order intraclasts, ooids, fossils, pellets.

The rock names are built up from the abbreviations already given, in the order allochem, matrix or cement, for example oosparite is made up of ooids in a sparry calcite cement. The abbreviation for rudaceous (rud) can be inserted for coarse varieties, for example a coarse intrasparite is an intrasparrudite.

Additional points to note are:

>10% replacement dolomite, add prefix 'dolomitised' to rock name. If the dolomite is of uncertain origin use the prefix 'dolomite'.

>10% terrigenous material, add prefix 'sandy', 'silty' or 'clayey' to rock name.

If one particular fossil type is dominant this can be shown in the rock name, e.g. crinoid biomicrite.

Dismicrite refers to disturbed micrite, the disturbance being due to bioturbation (which may lead to sparry calcite "birds eye") or physical action.

Biolithite refers to *in situ* organic build-ups.

Dunham Classification of limestones (Dunham, 1962)

The presence of mud-sized material ($<20\mu m$) is considered to be of sedimentological significance. As much of the carbonate material is produced essentially in the environment of accumulation, the feature of importance is how much fine material is allowed to remain at the site of deposition. Grains ($>20\mu m$) may be sparsely abundant in mud or they may be so abundant as to form a framework. The former are said to be 'mud-supported' while the latter are said to be 'grain-supported'. Although the majority of carbonate rocks are clearly made up of

Volumetric allochem composition		>10% allochems		<10% allochems		UNDISTURBED BIOHERM LIMESTONE TYPE IV
		sparry calcite cement > microcrystalline ooze matrix	microcrystalline ooze matrix > sparry calcite cement	MICROCRYSTALLINE LIMESTONES Type III		
		SPARRY ALLOCHEMICAL LIMESTONES TYPE I.	MICROCRYSTALLINE ALLOCHEMICAL LIMESTONES TYPE II	1-10% allochems	<1% allochems	
>25% intraclasts		intrasparite	intramicrite	intraclast-bearing micrite	micrite or dismicrite	biolithite
<25% intraclasts	>25% ooids	oösparite	oömicrite	oöid-bearing micrite		
	<25% ooids; volume ratio of fossils: pellets >3:1	biosparite	biomicrite	fossiliferous micrite		
	3:1-1:3	biopelsparite	biopelmicrite	pelletiferous micrite		
	<1:3	pelsparite	pelmicrite			

Fig. 2.17 – Classification of limestones (based on Folk, 1959, 1964; Type V replacement dolomites omitted).

original loose particles, in some cases binding by organic activity (for example, corals or algae) may be significant (boundstone).

The rock names used (Fig 2.18) are normally qualified by reference to the principal grain type, for example crinoid packstone, foraminiferal wackestone, etc.

Dunham classification modified and extended

Folk places all the autochthonous limestones in one rock group, biolithite. Dunham places them in boundstones. Embry and Klovan (1975) suggested division within this group.

Framestones contain *in situ* massive fossils which construct a rigid three-dimensional framework during deposition.

Bindstones contain *in situ* lamellar or tabular fossils which encrusted and bound the sediment during deposition. The supporting framework is the matrix.

Bafflestones contain *in situ* stalk-shaped fossils which during life trapped sediment by acting as baffles. This type of rock is subjective in its recognition.

This classification can be used to provide rock names, with information on the matrix type, for example thamnoporoid floatstone with a fine-grained, skeletal wackestone matrix.

In the case of autochthonous limestones the matrix is described on the >2 mm and <2 mm particle size scale, for example tabular stromatoporoid bindstone with a thamnoporoid floatstone matrix with a fine-grained wackestone matrix. The Folk and Dunham classifications ignore the matrix, for example tabular stromatoporoid biolithite or boundstone.

Depositional texture recognisable				Depositional texture not recognisable
Original components not bound together during deposition			Original components were bound together during deposition	Diagenetically recrystallised
Contains carbonate mud <20μm		no mud		
Mud-supported	Grain-supported			
<10% grains	>10% grains			
Mudstone	Wackestone	Packstone	Grainstone	Boundstone	Crystalline carbonate

Fig. 2.18 – Classification of limestones (after Dunham, 1962).

Organic Deposits

Allochthonous limestones						Autochthonous limestones		
Original components not bound organically during deposition						Original components organically bound during deposition		
less than 10% >2 mm components				greater than 10% >2 mm components		by organisms which act as baffles	by organisms which encrust and bind	by organisms which build a rigid framework
contains lime mud			no lime mud	Matrix supported	>2 mm grain supported			
mud-supported		grain-supported						
<10% grains (>0.03 to <2 mm)	>10% grains							
mudstone	wackestone	packstone	grainstone	floatstone	rudstone	bafflestone	bindstone	framestone

Fig. 2.19 — Classification of limestones (after Embry and Klovan, 1975).

Although complicated this classification gives a relatively complete description of the autochthonous limestone rock types.

Other limestone names
Many different systems of limestone nomenclature have been proposed but only those names which are commonly encountered in the literature are listed below.

Calcirudite – coarse grained (>2mm).

Calcarenite – sand-sized particles (0.06 – 2.0 mm) Equivalents of this are coquina and bahamite (with grapestone).

Calcisiltite – silt sized particles (0.004 – 0.06 mm). Otherwise called micrograined limestone.

Calcilutite – mud sized particles (<0.004 mm). Equivalents of this are chalk (coccolith micrite or nanno-ooze), calcite mudstone, aphanic limestone, lithographic limestone, cryptograined limestone.

Hybrid limestones
Some limestones contain admixtures of terrigenous material and this may be stated in the rock name, for example sandy biosparite – a biosparite containing quartz sand; clayey or argillaceous micrite.

Marl is a unlithified mixture of clay minerals and lime mud.

Marlstone is the lithified equivalent.

2.5.3 Dolomitic limestones and dolomites
The term *dolomite* is used both for the mineral and for rocks composed of >90% of the mineral although the name *dolostone* has been suggested for the latter.

In most cases the dolomite is the result of diagenetic replacement. Dolomitic limestones may still preserve enough of the original limestone fabric to be identifiable according to the Folk or Dunham schemes. In such cases the rock would be a dolomitic biomicrite, dolomitic boundstone, etc. The calcitic dolomites may preserve 'ghosts' of the original fabric.

In some cases the dolomite is rhombic. Where rhombs, make point contact with one another and are separated by pores the texture is said to be *saccharoidal* or sucrosic (that is, sugary).

Patchy dolomitisation may create the appearance of a breccia – such rocks are called *pseudobreccia.*

Autoclastic breccia develops when diagenetic shrinkage causes brecciation. This may be recemented.

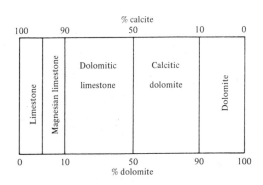

Fig. 2.20 – The intergradation of limestone and dolomite (after Pettijohn, 1957).

2.5.4 Siliceous sediments and rocks

Organic siliceous sediments may be rich in diatoms (unlithified, *diatomite;* lithified, *diatomaceous chert* or *diatomite)* in radiolarians (unlithified, *radiolarian ooze* lithified, *radiolarian chert* or *radiolarite).*

See also pp. 47, 48, deep sea sediments. Replacement cherts are considered under chemical precipitates, p.47.

2.6 CHEMICAL PRECIPITATES

2.6.1 Evaporites

The term *evaporite* refers to deposits of the following minerals:
Sulphates: anhydrite, gypsum and more rarely, kainite, kieserite, and polyhalite. Chlorides: halite, carnallite and sylvite.

2.6.2 Phosphorites

Apatite is the principal sedimentary phosphate mineral. Rocks with >20% apatite are said to be phosphatic; those with 50% apatite are called *phosphorites.*

2.6.3 **Ironstone**

Sedimentary rocks rich in iron are collectively called *ironstone.* Common iron-bearing minerals found in such sediments are pyrite, chamosite, siderite, limonite, and haematite.

2.6.4 **Caliche and calcrete**

Caliche is a carbonate development in the soil of semi-arid regions. The term calcrete is often used for indurated varieties.

Esteban (1976) gives the following discriptive definition: 'Caliche is a vertically zoned, subhorizontal carbonate deposit, normally developed with three main rock types:

(1) massive-chalky,
(2) nodular-crumbly,
(3) laminated and/or pisolitic compact crust or caprock'.

The main fabric is a 'clotted, pelloidal micrite with microspar channels and cracks' (Esteban, 1976). The thickness varies from centimetres to many metres (Reeves, 1970).

2.6.5 Silcrete
This is the siliceous equivalent of calcrete. Silcretes are characteristically brecciated. They are composed of chert and quartz.

2.6.6 Chert and flint
Chert is cryptocrystalline silica occurring as nodules (often replacing sediment, especially limestone, or infilling pre-existing cavities such as animal burrows) or bedded. *Flint* is a variety name given to chert.

2.7 SPECIAL GROUPS

2.7.1 Deep sea sediments
The increased availability of deep sea sediments as a consequence of the Deep Sea Drilling Project led to the development of the following classification scheme (Weser, 1971; Lancelot, *et al,*. 1977):

% $CaCO_3$	Unconsolidated	Semilithified	Lithified
0-30	clay	clay	claystone
30-60	marl	marl	marlstone
60-100	ooze	chalk	limestone

The sediment takes its name from those constituents present in amounts >25%. These are arranged in order of increasing abundance from left to right. For example a sediment made up of 30% zeolite and 70% coccoliths would be a zeolitic nanno ooze.

A minor constituent present to the extent of 10 - 25% is prefixed to the sediment name by the term rich, for example 10% zeolites, 30% radiolarians, 60% coccoliths = zeolite rich rad nanno ooze.

List of biogenic components and abbreviations () used

Calcareous

Nannofossil (nanno) – test of coccolithophorids, discoasters, etc.
Foraminiferal (foram) – principally planktonic forms Pteropod.
N. B. The term 'Globigerina ooze', much used in the past, would lapse in favour of foraminiferal ooze.

Siliceous

Radiolaria (rad)
Silicoflagellates
Diatoms
N. B. Sediments made up largely of radiolaria are termed 'radiolarites' and those of diatoms 'diatomites'. Amorphous varieties are termed chert or porcellanite.

2.7.2 Process name rock types

Some sedimentologists use a system of nomenclature which does not define the type of rock but the process thought to have given rise to it. Some examples are:

aeolianite (or eolianite) – a wind deposited sediment.
contourite – a deep sea sediment deposited by a contour current (Stow and Lovell, 1979).
laminite – a sediment showing laminae due to rhythmic changes in depositional conditions.
loessite – indurated wind-blown silt of periglacial origin (Edwards, 1979).
rhythmite – same as laminite (Reineck and Singh, 1975).
tempestite – material deposited by a storm (or tempest)
tidalite – intertidally deposited sediment, that is, indicating the existence of tides at the time of deposition.
turbidite – material deposited from a subaqueous turbidity current.

Such terms would be better employed as adjectives alongside a descriptive rock name, for example, turbiditic sandstone, as recognition of process is subjective.

2.8 REFERENCES

Bathurst, R. 1971 Carbonate sediments and their diagenesis. *Developments in Sedimentology,* **12**, 620p.

Cummins, W.A. 1962 The greywacke problem. *Lpool Manchr geol. J.,* **3**, 51-72.

Dott, R.H. 1964 Wacke, graywacke and matrix – what approach to immature sandstone classification? *J. sedim. Petrol.,* **34**, 625-632.

Dunham, R.J. 1962 Classification of carbonate rocks according to depositional texture. *Mem. Amer. Ass. Petrol. Geol.,* **1**, 108-121.

Edwards, M.R. 1979 Late Precambrian glacial loessite from North Norway and Svalbard. *J. sedim. Petrol.,* **49**, 85-92.

Embry, A.F. and Klovan, J.E. 1971 A late Devonian reef tract on northeastern Banks Island, Northwest Territories. *Bull. Can. Petrol. Geol.,* **19**, 730-781.

Esteban, M. 1976 Vadose pisolite and calciche. *Bull. Amer. Ass. Petrol. Geol.,* **60**, 2048-2057.

Fisher, R.V. 1961 Proposed classification of volcaniclastic sediments and rocks. *Bull. Geol. Soc. Am.,* **72**, 1409-1414.

Fisher, R.V. 1966 Rocks composed of volcanic fragments and their classification. *Earth-Sci. Rev.,* **1**, 287-298.

Folk, R.L. 1951 Stages of textural maturity in sedimentary rocks. *J. sedim. Petrol.,* **21**, 127-130.

Folk, R.L. 1959 Practical petrographic classification of limestones. *Bull. Amer. Ass. Petrol. Geol.,* **43**, 1-38.

Folk, R.L. 1962 Spectral subdivision of limestone types. *Mem. Amer. Ass. Petrol. Geol.,* **1**, 62-84.

Grim, R.E. and Güven, N. 1978 Bentonites. *Developments in Sedimentology,* **24**, 256pp.

Harland, W.B., Herod, K.N and Krinsley, D.H. 1966 The definition and identification of tills and tillites. *Earth-Sci. Rev.,* **2**, 225-256.

Honnorez, J. and Kirst, P. 1976 Submarine basaltic volcanism: morphometric parameters for discriminating hyaloclastites from hyalotuffs. *Bull. volcan.,* **39**, 441-465.

Imbrie, J. and Purdy, E.G. 1962 Classification of modern Bahamian carbonate sediments. *Mem. Amer. Ass. Petrol. Geol.,* **1**, 253-272.

Krumbein, W.C. 1934 Size frequency distributions of sediments. *J. sedim. Petrol.,* **4**, 65-77.

Krumbein, W.C. 1936 The application of logarithmic moments to size frequency distributions of sediments. *J. sedim. Petrol.,* **6**, 35-47.

Krynine, P.D. 1948 The megascopic study and field classification of sedimentary rocks. *J. Geol.,* **56**, 130-165.

Lancelot, Y., Seibold, E., and Gardener, J.V. 1977 Introduction. *Initial Repts. Deep Sea Drilling Project,* **41**, 7-18.

Lees, A. 1975 Possible influence of salinity and temperature on modern shelf carbonate sedimentation. *Mar. Geol.,* **19**, 159-198.

Lewan, M.D. 1978 Laboratory classification of very fine grained sedimentary rocks. *Geology,* **6**, 745-748.

Moore, L.R. 1968 Cannel coals, bogheads and oil shales. In Murchison, D. and Westoll, T.S. *Coal and coal-bearing strata.* Oliver & Boyd, Edinburgh, pp. 19-29.

Peltz, S. 1971 Quelques considérations sur la nomenclature et classification des pyroclastites. *Bull. volcan.,* **35**, 295-302.

Pettijohn, F.J. 1957 *Sedimentary rocks.* Second edition. Harper & Brothers, New York. 718pp.

Pettijohn, F.J. 1975 *Sedimentary rocks.* Third edition. Harper & Row, New York, 628pp.

Pettijohn, F.J., Potter, P.E. and Siever, R. 1972 *Sand and sandstone.* Springer – Verlag, Berlin. 618pp.

Reeves, C.C. 1970 Origin, classification, and geologic history of caliche on the southern High Plains, Texas and eastern New Mexico. *J. Geol.,* **78**, 352-362.

Reineck, H.E. and Singh, I.B. 1975 *Depositional sedimentary environments.* Springer – Verlag, Berlin pp. 439.

Schermerhorn, L.J.G. 1966 Terminology of mixed coarse-fine sediments. *J. sedim. Petrol.,* **36,** 831-835.

Selley, R.C. 1976 *An introduction to sedimentology.* Academic Press, London. 408pp.

Sparks, R.S.J. and Walker, G.P.L. 1973 The ground surge deposit: a third type of pyroclastic rock. *Nature, Phys. Sci.,* **241,** 62-64.

Stow, D.A.V. and Lovell, J.P.B. 1979 Contourites: their recognition in modern and ancient sediments. *Earth-Sci. Rev.,* **14,** 251-291.

Teichmüller, M. and Teichmüller, R. 1968 Geological aspects of coal metamorphism. In Murchison, D. and Westoll, T.S. *Coal and coal bearing strata.* Oliver & Boyd, Edinburgh, pp. 233-267.

Wagner, C.W. and van der Togt, C., 1973 Holocene sediment types and their distribution in the southern Persian Gulf. In Purser, B.H. (editor),*The Persian Gulf,* Springer – Verlag, Berlin, pp. 123-155.

Wentworth, C.K. 1922 A scale of grade class terms for clastic sediments. *J. Geol.* **30,** 377-392.

Wentworth, C.K. and Williams, H. 1932 The classification and terminology of the pyroclastic rocks. *Bull. National Res. Council.,* **89,** 19-53.

Weser, O.E. 1971 Sediment classification. *Initial Repts. Deep Sea Drilling Project,* **18,** 9-10.

Wright, A.E. and Bowes, D.R. 1963 Classification of volcanic breccias: a discussion. *Bull. geol. Soc. Amer.,* **74,** 79-86.

Chapter 3

Igneous Rocks

Igneous rocks may be classified on their texture, crystal or grain size, colour, mineralogy, chemical composition, mode of occurrence, and genesis. Classification systems based on chemical composition alone involve the recalculation of chemical data to give the equivalent 'mineralogical' composition or *norm* (for Niggli values and the Cross, Iddings, Pirsson and Washington, that is, C.I.P.W., method see Hatch, Wells and Wells, 1972, or Nockolds, Knox and Chinner, 1978). Such systems are not appliclicable to field use as the chemical analysis has to be carried out in order to classify the rock. They also ignore the mineralogical and texture features considered to be important by petrographers. The commonly used classification schemes are based primarily on the content of essential minerals, that is, those that make up the bulk of the rock, known as the *mode* of the rock (see Hatch, Wells and Wells, 1972). Minor amounts of accessory minerals are not taken into consideration.

The development of the classification of igneous rocks has a long history and has involved many distinguished petrographers, amongst whom may be mentioned Harker, Teall, Flett, Bailey, Holmes, A. K. and M. K. Wells in Britain, and Johannsen, Rosenbusch, Iddings, Chayes, from abroad. However, difficulties have arisen because different authors have used the same name for rocks of slightly different mineral composition.

In order to overcome these problems and to produce a classification scheme which was standardised and internationally acceptable, Streckeisen (1964, 1965, 1967) produced a series of papers outlining classification schemes for comment and debate by petrologists. The discussion meeting which was to have been held at the International Geological Congress in Prague in 1968 could not be held. Following this, the International Union of Geological Sciences set up a Subcommission on the Systematics of Igneous Rocks under the chairmanship of Professor Streckeisen. The classifications proposed by this Subcommission are reproduced here.

The Subcommission concluded "By igneous we mean as far as classification and nomenclature are concerned, 'massige Gesteine' in the sense of Rosenbusch or 'igneous and igneous-looking rocks' of Anglo-Saxon authors, irrespective of their genesis. They may have crystallised from magmas (including cumulates) or may have come into being by deuteric, metasomatic, or metamorphic processes. We should have learned from the granite controversy that phaneritic rocks can form by multiple processes, and it would be unwise to give the same rock different names according to the origin that is assumed by different authors" (Streckeisen, 1976, p.4).

The aim was to make the naming and classification of igneous rocks more precise. To this end they used names defined on mineral content and texture, accepting only those names in common usage, and rejecting names based on rock chemistry, those of local application only and those with a genetic connotation. The schemes proposed represent a compromise between the opinions of different individuals and different countries and it remains to be seen whether they will be widely accepted. "It is to be hoped that the proposals and the others which will follow will be used generally yet will not oversimplify the petrographical relationship of complicated mineral assemblages that rocks represent, nor stultify the natural development of a subject where pigeon-holes are a necessary evil that should not be obtrusive" (Sabine, 1974, p.174).

Igneous rocks are commonly divided into three main groups on the basis of their field relationships or on their average grain or crystal size:

plutonic – crystallised at depth; average crystal size >5 mm (coarse) or 1 – 5 mm (medium)

hypabyssal – crystallised at shallow depth; average crystal size 0.5 – 1 mm (fine grained)

volcanic – extruded at the surface of the earth; average crystal size ≤0.5 mm (very fine grained)

Clearly there are many instances where the mode of occurrence is uncertain so commonly greater emphasis is placed on the crystal size. However, even these divisions are used with some flexibility, for example where different parts of the same rock have different average crystal sizes.

3.1. **PLUTONIC ROCKS**

Plutonic rocks are classified according to their modal mineral content (that is, actual mineral content measured in volume percent). A primary division is made on the mafic and related mineral content (M). Where M is <90% the rocks are classified on their light coloured mineral content. Where M is 90 – 100% the rocks are classified according to their mafic minerals. M includes micas, amphiboles, pyroxenes, olivines, opaque minerals, zircon, apatite, epidote, garnets, etc.

3.1.1. $M < 90\%$

These rocks are classified on the basis of their position on the Q A P F double triangle. The mineral groups are:

Q quartz

A alkali feldspar – orthoclase, microcline, perthite, anorthoclase, albite An_{00-05}

P plagioclase An_{05-100}, scapolite

F feldspathoids or foids – leucite, pseudoleucite; nepheline, sodalite, nosean, hauyne, cancrinite, analcite, etc.

These light coloured minerals are calculated to 100% ignoring the mafic content.

$Q + A + P = 100$ or $A + P + F = 100$.

Thus, only in those rocks made up wholly of felsic minerals is percent Q the same as the modal value for quartz (see Hatch, *et al*, 1972, p.196). Precise identification of plutonic igneous rocks is based on microscopic examination of thin sections. However, the broad groups can be recognised in the field, and a simple classification for field use is presented in Figs 3.1 to 3.3. Nevertheless, many geologists do not use this terminology as they prefer to use granite, basalt, etc. in an informal way.

An outline of the laboratory classification of plutonic rocks is given in Figs 3.4 to 3.9. To the root names listed in Fig 3.4, prefixes may be added to describe the mineralogy, the mineral names being listed in order of increasing abundance, for example, hornblende-biotite granodiorite, with some hornblende but more biotite.

3.1.2 $M > 90\%$ – ultramafic rocks

A simple division into peridotite (40-100% olivine), pyroxenite (<40% olivine, >50% pyroxene) and hornblendite (<40% olivine, >50% hornblende) can be further subdivided into varieties (Fig. 3.8). Ultramafic rocks lacking hornblende may be divided as shown in Fig. 3.9.

Plutonic Rocks

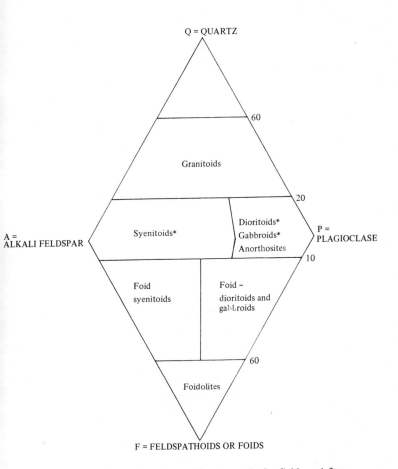

Fig. 3.1 — General classification of igneous rocks for field use (after Streckeisen, 1974). *Add 'foid-bearing' if foids are present.

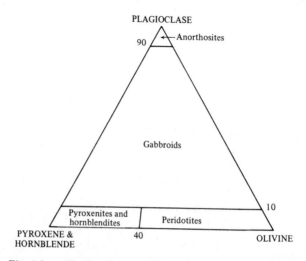

Fig. 3.2 — Classification of gabbroic and ultramafic rocks for field use (after Streckeisen, 1974).

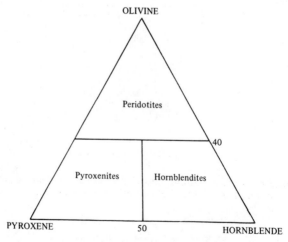

Fig. 3.3 — Classification of ultramafic rocks for field use (after Streckeisen, 1974).

Plutonic Rocks

CLASS	ORDER	DIVISION		
		PLUTONIC	M	VOLCANIC
M <90	I Quartz rocks	1a. Quartzolite 1b. Quartz-rich granitoids		
	II Quartz-feldspar rocks	2. Alkali-feldspar granite 3. Granite 4. Granodiorite 5. Tonalite	0-20 5-20 5-25 10-40	Alkali-feldspar rhyolite Rhyolite Dacite Dacite
	III Feldspar rocks	6. Alkali-feldspar syenite	0-25	Alkali-feldspar trachyte
		7. Syenite	10-35	Trachyte
		8. Monzonite	15-45	Latite
		9. Monzodiorite and Monzogabbro	20-50 25-60	Andesite
		10. Diorite Gabbro Anorthosite	25-50 35-65 0-10	Basalt
	IV Feldspar - feldspathoidal rocks	11. Foid syenite	0-30*	Phonolite
		12. Foid monzosyenite	15-45	Tephritic phonolite
		13. Foid monzodiorite, foid monzogabbro	20-60	Phonolitic tephrite
		14. Foid diorite, foid gabbro	30-70	Tephrite or basanite
	Feldspathoidal rocks	15. Foidolites		Foidite
M >90	VI Ultramafic rocks	16. Peridotite Pyroxenite Hornblendite		Ultramafitite

Fig. 3.4 — Classification of igneous rocks (modified from Streckeisen, 1967). M values are listed as normal; the prefixes 'leuco' and 'mela' may be added to the rock name where the values are respectively lower or higher than the normal range; *30 - 60 = malignite, 60 - 90 = shonkinite.

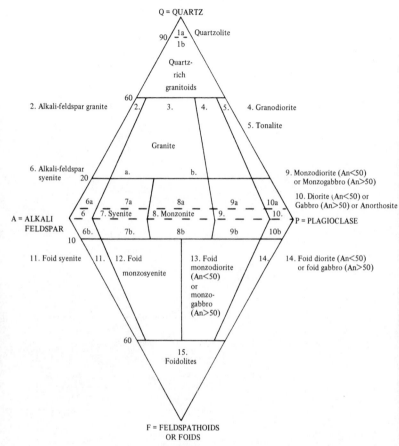

Fig. 3.5 – General classification of plutonic rocks with M<90 (after Streckeisen, 1974). For groups 6 - 10, a = quartz, e.g. 6a is Alkali feldspar quartz syenite, and b = foid-bearing, e.g. 6b is foid-bearing alkali feldspar syenite.

Notes on Fig. 3.5. Numbers refer to fields.

2. *Alkali-feldspar granites.* The nature of the feldspar present should be indicated in the rock name, for

example, albite granite. When alkali amphibole or alkali pyroxene (i.e. soda-rich) are present. The term 'alkali granite' is used.

3. *Granite.* Some authors prefer to divide this field according to the proportions of alkali-feldspar versus plagioclase. The Subcommision argues that the term granite should be applied to the most commonly occurring granitic rock. These fall in field 3b which is termed by other authors adamellite, quartz-monzonite or monzogranite.

4. *Granodiorite.* This is a common rock type. Local varieties include opdalite, a dark, hypersthene - bearing granodiorite, and enderbite, charnockitic granodiorite.

5. *Tonalite.* Hornblende and/or biotite may be present. Light coloured tonalites containing oligoclase or andesine are also called trondhjemite or plagiogranite.

6. *Alkali-feldspar syenite.* Where quartz forms 5-20% the rock is a quartz – bearing alkali syenite or nordmarkite. Where feldspathoids form 0-10% the rock is a feldspathoid-bearing alkali syenite or pulaskite.

7. *Syenite.* Otherwise called calc-alkaline syenite.

8. *Monzonite.* Specific variety names are kentallenite for dark olivine monzonite, and larvikite for those varieties with large crystals of oligoclase (antiperthitic) and alkali feldspar.

10. *Diorite, gabbro, anorthosite.* Typical diorites are composed of andesine, hornblende and/or biotite, sometimes with augite but rarely with olivine. Typical gabbros (including norites) are composed of labradorite or bytownite, pyroxene (diallage and/or orthopyroxene), together with olivine quite frequently, and occasionally amphibole.

Gabbro contains clinopyroxene and norite contains

orthopyroxene. Rocks with >5% each of orthopyroxene and clinopyroxene are gabbronorites. For further subdivisions see Figs. 3.6 and 3.7.

Rocks transitional between gabbro and diorite are termed gabbro-diorite.

Anorthosite generally contains labradorite or andesine.

Alkaline rocks. This group includes all those that contain feldspathoids with or without alkali pyroxenes and/or alkali amphiboles, for example, the A-P-F triangles of fig. 3.5. Not all the criteria necessary for the naming of these rocks are presented in this figure. The nature of the feldspathoids and mafic minerals, colour index, and textural relationships must be taken into account. The feldspathoid should be stated in the name of all rocks from fields 11 - 15, for example, nepheline syenite.

12. *Foid monzosyenite.* An alternative root name is plagisyenite.
13. *Foid monzodiorite, foid monzogabbro.* A synonym is essexite.
14. *Foid diorite, foid gabbro.* Nepheline gabbro is also called theralite. Teschenite is analcime gabbro.

Plutonic Rocks

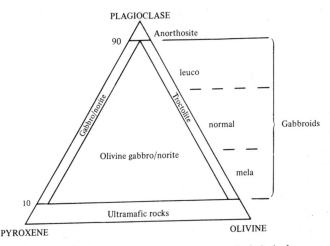

Fig. 3.6 — Classification of gabbroic rocks composed of plagioclase, pyroxene and olivine (after Streckeisen, 1974).

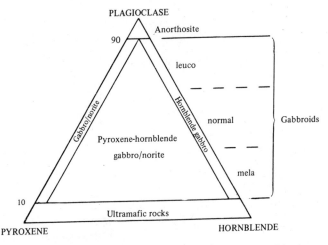

Fig. 3.7 — Classification of gabbroic rocks containing hornblende (after Streckeisen, 1974).

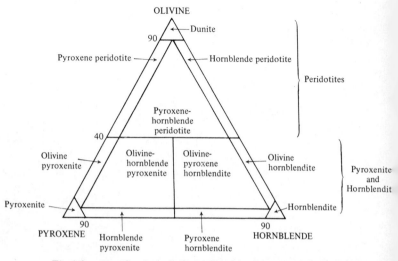

Fig. 3.8 — Classification of ultramafic rocks that contain hornblende (after Streckeisen, 1974).

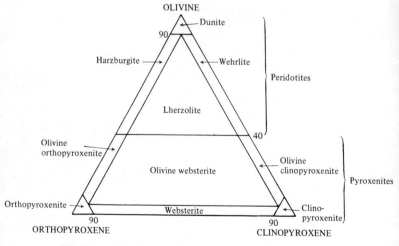

Fig. 3.9 — Classification of ultramafic rocks composed of olivine, orthopyroxene and clinopyroxene (after Streckeisen, 1974).

Where garnet or spinel is present it may be mentioned in the name, for example, <5% garnet in peridotite = garnet-bearing peridotite; >5% spinel in pyroxenite = spinel pyroxenite.

Common varieties: orthopyroxenite (bronzitite, enstatitite, hypersthenite); clinopyroxenite (diopsidite, diallagite); dunite (olivinite). Komatiite: magnesium – rich ultramafic rocks. Some show affinites with peridotites and others with tholeiitic basalts (Viljoen and Viljoen, 1970).

3.1.3 Charnockites

Charnockites are regarded as igneous or igneous-looking, and they are characterised by the presence of hypersthene and perthitic feldspars. The Subcommision recommends using the prefix 'hypersthene' with the appropriate rock name although special terms are also used.

Field	General terms	Special terms
2	hypersthene alkali-feldspar granite	alkali-feldspar charnockite
3	hypersthene granite	charnockite
4	hypersthene granodiorite	opalite
5	hypersthene tonalite	enderbite
6	hypersthene alkali-feldspar syenite	
7	hypersthene syenite	
8	hypersthene monzonite	mangerite
9	hypersthene monzodiorite or monzonorite	jotunite
10	norite (hypersthene diorite)	

Charnockite rock name (from Streckeisen, 1976).

3.2 HYPABYSSAL ROCKS

These crystallise at high crustal levels and form minor intrusive bodies such as dykes, sills, etc. Commonly they are fine grained varieties of plutonic rocks. Those lacking phenocrysts are said to be aphyric while those with phenocrysts are termed porphyritic (The term porphyrite has fallen out of favour, Sabine, 1978).

The prefix 'micro' is used with the appropriate plutonic rock name, for example, aphyric microgranite, porphyritic microgranite.

In the case of dykes and other sheet-like intrusions fine grained varieties are said to be aplites and coarse grained varieties pegmatites e.g. granite aplite, granite pegmatite.

There are some names specifically for hypabyssal rocks:

granophyre – alkali-feldspar granite and granite having a micro-pegmatitic texture.

dolerite, diabase – gabbroic rocks with subophitic or ophitic texture (plagioclase laths partly or wholly enclosed in pyroxene). Some contain quartz, others olivine. Formerly diabase was used as a term for altered basalts but this usage is not now favoured (Sabine, 1978; Streckeisen, 1979). The term *greenstone* has also in the past been used for altered dolerites.

teschenite – foid monzonite, foid monzogabbro, foid diorite, foid gabbro.

picrite – olivine and pyroxene, sometimes with biotite and amphibole.

lamprophyre – dark rocks forming dykes. They normally contain biotite or hornblende together with augite (see Fig. 3.10).

carbonatites – these may be intrusive or extrusive. Carbonate minerals make up $\geq 50\%$ by volume. The type of carbonate mineral is used to define four classes (Sabine, 1978; Streckeisen, 1979).

1. Calcite carbonatite:
   sövite – coarse grained, light coloured, 90% calcite.

FELSIC CONSTITUENTS		PREDOMINANT	MAFIC	MINERALS
Feldspar	Foid	Biotite, diopside, augite, ± olivine	Hornblende, diopside, augite, ± olivine	Amphibole (barkevikite, kaersutite), titanaugite, olivine, biotite	Mellite, biotite, ± titanaugite, ± olivine, ± calcite
Colour index		>35	>35	>40	>70
or > pl		Minette	Vogesite	Sannaite	Polzenite
pl > or		Kersantite	Spessartite	Camptonite	Alnoite
or > pl	fsp > foid			Monchiquite	
pl > or	fsp > foid				
absent	glass or foid				
absent	absent				
LAMPROPHYRE TYPE		calc – alkaline		alkaline	melilitic

Fig. 3.10 – Lamprophyre mineral compositions (after Streckeisen, 1979).

alvikite – medium to fine grained, light coloured, calcite.

2. Dolomite – carbonatite:
   beforsite – >90% dolomite
3. Ferrocarbonatite – iron-rich carbonates.
4. Natrocarbonatite – alkali (Na and K) and calcium carbonates.

3.3 VOLCANIC ROCKS

Volcanic rocks are commonly fine-grained or even glassy so that modal mineral composition cannot always be determined. For those cases where it is possible the schemes outlines in Figs 3.11 and 3.12 are proposed (Sabine, 1978, Streckeisen, 1979). In other cases chemical composition has to be considered as a basis for classification, but direct correlation of mineral and chemical classifications is difficult.

The general classification shown in Fig. 3.12 gives the root-names only, and these are arranged and numbered as in the classification of plutonic igneous rocks (Fig. 3.5).

2. Alkali feldspar rhyolite (liparite): the name used for those volcanic rocks with alkali pyroxene and/or alkali amphibole in the mode or in the norm is alkali rhyolite. Pantellerite contains soda pyroxene.

3b. The term rhyodacite may be used for rocks intermediate between rhyolite and dacite.

6-8. Rocks in these fields which contain no modal foid but contain nepheline is the norm may be said to be 'ne-normative'.

9,10 The majority of volcanic rocks fall in these fields. Supplementary criteria are neccessary for their subdivision. Various criteria have been used for the

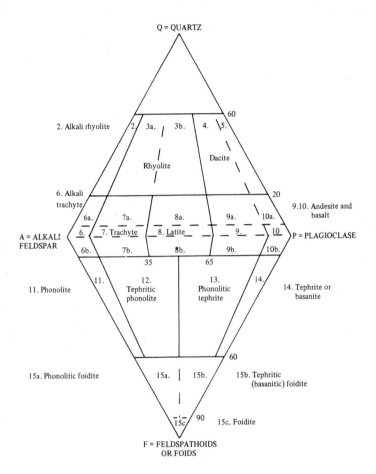

Fig. 3.11 — General classification of non-ultramafic volcanic igneous rocks (after Sabine, 1978, and Streckeisen, 1979). For 6a - 8a add the prefix 'Quartz'; for 6b - 8b add the prefix 'foid-bearing'.

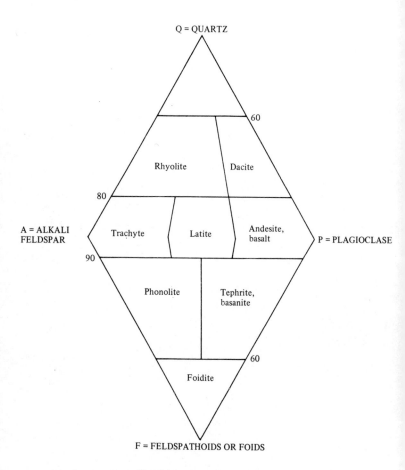

Fig. 3.12 — Simplified classification of non-ultramafic volcanic igneous rocks (after Sabine, 1978, and Streckiesen, 1979).

distinction between andesite and basalt. The Subcommision recommends using the silica content and colour index (see Fig. 3.13).

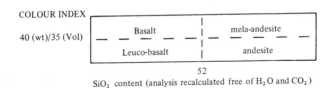

Fig. 3.13 — Distinction between basalt and andesite (after Streckeisen 1979).

Calc-alkaline andesites (with clinopyroxene) are mainly distributed in field 9a. Calc-alkaline basalts are commonly restricted to field 10. Orthopyroxene is the main ferromagnesian component. Tholeiitic basalts (that is, poor or lacking in olivine) occupy fields 10 and 10a. The pyroxenes are subcalcic augite and/or pigeonite.

Alkali basalts, having titanaugite and olivine but only a small feldspathoid content, occupy field 10b, and hawaiites, with andesine and anorthite in equal amounts, are mainly confined to this field too. Mugearites, having anorthite as the dominant feldspar, are mainly confined to fields 9 and 9b.

11. Phonolite has alkali feldspar and any feldspathoid, usually also with mafic minerals. If feldspathoid is present as a major constituent it must be indicated in the name of the rock, for example, leucite phonolite.

13,14 The name basanite applies where the olivine content is > 10%.

15c. The root name is foidite and these rocks may be named according to their dominant feldspathoid.

nepheline – nephelinite
olivine nephelinite (='nepheline basalt')
leucite – leucitite
olivine leucitite (='leucite basalt')
analcime – analcimite
olivine analcimite ('analcime basalt').

The use of the term 'basalt' for these rocks is considered incorrect as the recommended usage is for rocks containing feldspar.

16. Ultramafic volcanic rocks. The root is ultramafitite which may be qualified by the predominant mafic mineral, for example, melilite ultramafitite, or melilitite.

3.3.1 Glassy rocks.

The glass content (in % by volume) of volcanic rocks may be indicated thus:

0 – 20%	glass-bearing
20 – 50%	glass-rich
50 – 80%	glassy
80 –100%	obsidian, pitchstone

Porphyritic rocks with a glassy matrix may be named according to their chemical composition with the prefix 'hyalo', for example, hyalo-rhyolite (formerly called vitrophyre).

Basaltic glass formed at chilled margins is called tachylite if fresh and palagonite if altered.

3.3.2. Igneous rock series

From the point of view of genesis three rock series are recognised. In each there is a broad series of rock types with elements in common (Nockolds, Knox and Chinner, 1978).

Tholeiitic igneous rock series – relatively rich in iron, for example, tholeiitic basalts, tholeiitic andesites, ferro varieties of dacites, rhyolites, diorites, granodiorites and granites, many gabbros and some ultrabasic rocks.

Calc-alkali igneous rock series – relatively poor in iron and generally richer in alumina than rocks of the previous series, for example, calc-alkali basalts, andesites, latite-andesites, dacites, rhyolites, some gabbros and ultramafic rocks, diorites, monzodiorites, tonalites, grandiorites and granites.

Alkali igneous rock series – in comparison with the above two series these are characterised by a higher content of alkalies for any given silica content. Mildly alkaline types: alkali basalts, hawaiites and mugearites, alkali gabbros, alkali monzodiorites and alkali monzonites. Highly alkaline types: basanites, foid monzogabbro, teschenites and ultra-alkaline rocks. Felsic types include phonolites and nepheline syenites, peralkaline (that is, undersaturated with alumina leading to the formation of soda-pyroxene and soda-amphibole) and alkali varieties of trachytes, syenites, rhyolites and granites.

3.3.3 Metamorphosed volcanic rocks

Volcanic rocks which have undergaone alteration but still preserve recognisable igneous texture may bear the prefix 'meta-', for example, metabasalt. In the past the terms diabase and greenstone were used to denote altered basalts but this is now not recommended.

Spilite is a basaltic rock with the texture and fabric of an eruptive rock but the albite – chlorite mineral association may be due to metamorphic or metasomatic changes. A typical mode of occurrence is as a submarine lava flow, often with pillow structure. Commonly associated with spilites are keratophyre and quartz keratophyre.

Keratophyre is a leucocratic intermediate volcanic rock with albite phenocrysts in a matrix of albite and accessory potassium feldspar, mafic minerals, quartz, calcite, etc. Quartz keratophyre contains quartz in the matrix and sometimes also as phenocrysts.

3.4 REFERENCES

Hatch, F. H., Wells, A. K. and Wells, M. K. 1972. *Petrology of igneous rocks.* 13th Edition, Murby. pp. 551.

Nockolds, S. R., Knox, R. W. O'B, and Chinner, G. A. 1978. *Petrology for students.* Cambridge University Press, pp. 435.

Sabine, P. A. 1974 How should rocks be named? *Geol. Mag.* **111**, 165-176.

Sabine, P. A. 1978 Progress on the nomenclature of volcanic rocks, carbonatites, melilite-rocks and lamprophyres. *Geol. Mag.* **115**, 463-466.

Streckeisen, A. 1965 Die Klassifikation der Eruptivgestein (Ergebnis einer Umfrage). *Geol. Rdsch.* **55**, 478-491.

Streckeisen, A. 1964 Zur Klassifikation der Eruptivgesteine. *N. Jb. Miner. Mh.* 195-222.

Streckeisen, A. L. 1967 Classification and nomenclature of igneous rocks (Final report of an inquiry). *N. Jb. Miner., Abh.* **107**, 144-466.

Streckeisen, A. 1974 Classification and nomenclature of plutonic rocks. *Geol. Rdsch.* **63**, 773-786.

Streckeisen, A. L. 1976 To each plutonic rock its proper name. *Earth Sci. Rev.* **12**, 1-33.

Streckeisen, A. 1979 Classification and nomenclature of volcanic rocks, lamprophyres, carbonatites, and melilitic rocks: recommendations and suggestions of the IUGS Subcommision on the Systematics of Igneous Rocks. *Geology,* 7, 331-335.

Vijoen, M. J. and Viljoen, R. P. 1970 The geology and geochemistry of the lower ultramafic unit of the Onverwacht Group and a proposed new class of igneous rocks. *Geol. Soc. S. Africa. Spec. Publ.* 2, 55-85.

Chapter 4

Metamorphic Rocks

> Metamorphism is the mineralogical and structural (textural) adjustment of (dominantly) solid rock to physical and chemical conditions which differ from those under which the rocks originated. Weathering and similar processes are conventionally excluded. (Spry, 1969, p.1).

The processes of diagenesis merge with those of metamorphism. Diagenesis is considered to take place at temperatures not very much greater than those of deposition of sediment. Metamorphism commences at higher temperatures than those of deposition (~150°C Miyashiro, 1973). The boundary between diagenesis and metamorphism should be defined by definite mineral assemblages. As regards the upper temperature limit of metamorphism Miyashiro (1973, p.21) suggests that "... metamorphism should be defined as processes taking place in essentirally solid rocks which are roughly below the temperature of the beginning of melting. The possible existence of a small proportion of silicate melt or aqueous fluid in the interstices between mineral grains, however, is not prohibited."

The important controls are temperature, confining pressure, chemical activity of water, deformation, and time. The relative value of these controls determines the nature of the metamorphism:

Thermal or contact metamorphism: heat dominant. Rocks adjacent to an intrusive igneous body recrystallise due to the rise in temperature.

Cataclastic metamorphism: crushing and grinding of rocks due to movement on faults. The temperature is too low to cause recrystallisation.

Regional or orogenic metamorphism.
Regional metamorphism takes place during an orogeny involving convergent lithospheric plates. Regionally metamorphosed rocks occur as extensive belts, "... hundreds or thousands of kilometres long and tens or hundreds of kilometres wide, within an ancient orogenic belt on a continent or on an island arc." (Miyashiro, 1973, p.24). There is variation in the amount of deformation and the temperatures involved. The processes involved take place over a long period and include successive phases of recrystallisation and deformation. Areas of regionally metamorphosed rocks are also commonly the sites of intrusion of igneous bodies.

Ocean floor metamorphism.
Relatively high heat flow along mid-ocean ridges is thought to cause metamorphism of basic and ultrabasic rocks beneath the basaltic layer (Miyashiro, 1973).

Metamorphic rocks are not arranged in any formal classification. The rock names are based mainly on texture and to a lesser extent on composition. In cases where the parent rock type has undergone limited metamorphism so that it is clearly identifiable the prefix 'meta' – is commonly used, e.g. metabasalt, metagreywacke, etc.

4.1 NAMES BASED ON TEXTURE

Slate: a fine grained rock with slaty cleavage.
Phyllite: a fine grained rock with schistosity due to the development of microsopic mica and chlorite.

Schist: a foliated rock, rich in mica flakes, which are visible to the naked eye, generally fairly homogeneous.

Gneiss: a medium to coarse grained rock rich in feldspar (often with quartz and mica) with bands of different mineral composition and colour. (see below).

Fels or rock: massive metamorphic rocks without schistosity.

Hornfels: a fine-grained, flinty rock, lacking preferred orientation of grains, often with porphyroblasts giving a spotted appearance.

Mylonite: 50-90% fine-grained to crypto-crystalline, commonly foliated matrix with strained porphyroclasts, that is, fragments of crystals. (see Fig.4.1).

Higgins (1971) terms the foliation 'fluxion structure' meaning flow structure.

Cataclasite: similar to mylonite but with an unfoliated matrix (see Fig. 4.1).

Buchite: glass produced by partial fusion of country rock adjacent to an intrusion.

Where appropriate the textural name is preceeded by mineral names, for example, staurolite schist.

4.2. NAMES BASED ON COMPOSITION

Quartzite or metaquartzite: a recrystallised quartzose rock with sutured grain boundaries.

Marble: a recrystallised calcareous rock (calcite and/or dolomite) with sutured or polygonal grain boundaries.

Skarn: a calc-silicate rock, normally the product of metasomatism of an impure limestone.

Serpentinite: a rock composed mainly of serpentine with some talc and chlorite.

Amphibolite: a hornblende-plagioclase rock. The hornblende is commonly aligned to give a lineation. Sometimes such rocks are called hornblende schists. If

NATURE OF MATRIX		PROPORTION OF MATRIX			
		0 – 10%	10 – 50%	50 – 90%	90 – 100%
Crushed	Foliated	Crush (tectonic) breccia or conglomerate	Protomylonite	Mylonite	Ultramylonite
Crushed	Massive	Crush (tectonic) breccia or conglomerate	Protocataclasite	Cataclasite	Ultracataclasite
Recrystallised	Minor	Hartschiefer			
Recrystallised	Major	Blastomylonite			
Glassy		Pseudotachylite or Hyalomylonite			

Fig. 4.1 – Classification of dynamically metamorphosed rocks (from Spry, 1969).

a relict ophitic texture is present the term epidiorite is commonly used.

Eclogite: omphacite (jadeite-diopside) and some garnet.

Granulite: in older usage it refered to rocks with a granular texture. Now it is used for rocks of granulite facies, that is, regional metamorphism under water deficient conditions. Hypersthene is a characteristic mineral in basic, intermediate and acid varieties, for example, charnockite, with quartz, orthoclase, plagioclase and hypersthene.

4.3 GNEISS

Spry (1969) lists the following textural varieties:

Granite (diorite, etc.) gneiss: massive, homogeneous, granoblastic to porphyroblastic, quartz-feldspar or feldspar rock. Mica occurs as separate parallel flakes and schlieren, and prisms of amphibole may be present.

Banded gneiss: megascopically banded with alternate granoblastic (quartz and feldspar) and schistose (mica) layers.

Augen gneiss: irregular lenticular banding with large eye shaped (augen) of feldspar.

Pencil gneiss: the fabric is dominated by lineation, due either to the intersection of co-axial foliations or to the presence of elongated mineral aggregates.

Conglomerate gneiss: the clastic texture is largely obliterated through recrystallisation.

Non-foliated gneiss: inhomogeneous, contorted rock, with a complex texture.

Gneisses derived from igneous rocks are called ortho-gneisses, while those derived from sedimentary rocks are para-gneisses.

Fig 4.2 and 4.3 show a classification of common, metamorphic rocks composed mainly of either quartz,

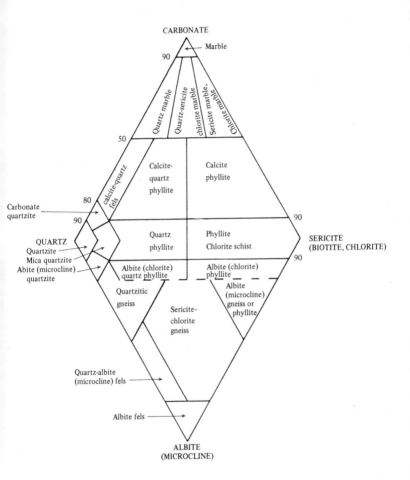

Fig. 4.2 — Classification of lower temperature range metamorphic rocks (from Winkler, 1974).

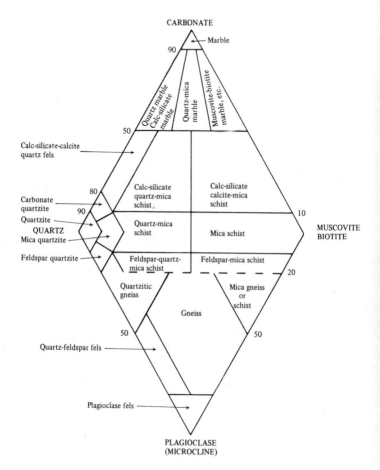

Fig. 4.3 — Classification of higher temperature range metamorphic rocks (from Winkler, 1974).

feldspar and phyllosilicates, or quartz, carbonates and phyllosilicates. In naming rocks the texture is the most important feature, and the boundaries between phyllite, schist and gneiss are more transitional than is implied in these diagrams. The order of mineral names preceeding the rock name is in increasing order of abundance, for example calcite-quartz phyllite, calcite < quartz.

4.4 REFERENCES

Higgins, M. 1971. Cataclastic rocks. *Prof. Pap. U. S. geol. Surv.*, **687**, 1-97.

Mason, R. 1978. *Petrology of the metamorphic rocks.* Allen and Unwin, London. pp. 254.

Miyashiro, A. 1973. *Metamorphism and metamorphic belts.* Allen and Unwin, London, pp. 492.

Spry, A. 1969. *Metamorphic textures.* Pergamon Press, pp. 350.

Winkler, H. G. F. 1979. *Petrogenesis of Metamorphic Rocks.* Fifth edition Springer-Verlag, Berlin, pp. 348.

Chapter 5

Mixed Rocks and Rock Associations

5.1 MIGMATITE

The term migmatite was introduced by Sederholm to embrace those rocks transitional between granites and crystalline schists. Migmatite refers to the mixed constituents. Usage of the word continues to be subject to controversy. The definition given below is that of Mehnert (1968, p. 355).

Migmatite – a megascopically composite rock consisting of two or more petrographically different parts, one is the country rock in a more or less metamorphic stage, the other is of pegmatitic, aplitic, granitic, or generally plutonic appearance.

The different types of penetration fabrics have been used to subdivide the migmatite group. Mehnert (1968) uses the following descriptive terms:

paleosome – the unaltered or only slightly modified parent or country rock.

neosome – the newly formed rock portion. Two rock types may be distinguished.

   leucosome – containing light-coloured minerals such as quartz and feldspar.
   melanosome – containing dark-coloured minerals such as biotite, hornblende, etc.

5.1.1 Structural types (after Mehnert, 1968)

Agmatic or breccia structure. Fragments of paleosome are enclosed by veins of neosome. The rock type is sometimes called agmatite. Brown (1973) regards this as an intrusion breccia rather than a migmatite.

Diktyonitic or net-like structure. Narrow veins of neosome interlace the paleosome.

Schollen or raft structure. The paleosome fragments are rounded and float like rafts in the neosome.

Phlebitic or vein structure. Tabular veins of neosome cut the paleosome in an irregular fashion. Such rocks are called veined gneiss.

Stromatic or layered structure. The neosomes form light and dark layers, generally parallel to the schistocity, within the paleosome.

Surreitic or dilation structure. Various structures developed in layered migmatites, where the mechanical strength of individual layers differ one from the other, for example boudins.

Folded structure. Various types of folds some of which are contemporaneous with the process of migmatisation while others predate this process.

Ptygmatic structure. Disharmonic folds in which there is no clear relationship between the thickness of the folded material and the limbs and axial regions of the folds.

Ophthalmic or augen structure. The neosome occurs as eye-shaped lenses within the palaeosome. This structure is very common in migmatites. The rock type is called augen gneiss.

Stictolithic or fleck structure. The mafic minerals are concentrated in flecks surrounded by light mafic-poor haloes.

Schlieren structure. Elongated streaks of light and dark minerals, caused by laminar flow.

Nebulitic structure. The paleosome and neosome are not separable except as ghost-like diffuse slight variations in mineral assemblages.

5.1.2. Metatexite and diatexite.

These two names were introduced to indicate rock genesis by metatexis and diatexis respectively. These processes represent degrees of anatexis, that is, re-melting or fusion. As with all genetic terms their use is controversial (*see* Brown, 1973).

5.2 OPHIOLITE

The term ophiolite has had a variety of different meanings during the past, but at a Penrose Conference of the Geological Society of America agreement was reached on the following definition (Coleman, 1977):

"Ophiolite ... refers to a distinctive assemblage of mafic to ultramafic rocks. It should not be used as a rock name or as a lithologic unit in mapping."

Where completely developed an ophiolite suite has the following sequence:

mafic volcanic complex – often with pillows

mafic sheeted dyke complex

gabbroic complex – with cumulus peridotites and pyroxenites and not so deformed as the ultramafic complex

ultramafic complex – harzburgite, lherzolite and dunite often serpentinised and with a metamorphic tectonic fabric.

The associated rock types include intrusive and extrusive sodic felsic rocks, chromite pods associated with the dunite, and overlying sediments which comprises cherts, shales and thin limestones.

Many authors believe ophiolites to represent oceanic crust sequences.

5.3 RODINGITE

This is a general term for metasomatised mafic rocks, the low temperature reaction zones being related to the process of serpentinisation. They occur within or in contact with serpentinite and are often involved in tectonic movements that have affected the serpentinite. Widely differing rocks types may be metasomatised to become rodingite. The characteristic mineral is hydrogarnet (see Coleman, 1977).

5.4 KIMBERLITE

"Kimberlite is serpentinized and carbonated mica-peridotite of porphyritic texture, containing nodules of ultrabasic rock-types characterized by such high-pressure minerals as pyrope and jadeitic diopside; it may, or may not contain diamond." (Dawson, in Wylie, 1967, p.242).

5.5 REFERENCES

Brown, M. 1973 The definition of metatexis, diatexis and migmatite. *Proc. Geol. Ass.* **84,** 371-382.

Coleman, R. G. 1977 *Ophiolites. Ancient oceanic lithosphere?* Springer – Verlag, Berlin. pp.229.

Mehnert, K. R. 1968 *Migmatites and the origin of granite rocks.* Elsevier, Amsterdam. pp. 393.

Wylie, P. J. Editor. 1967 *Ultramafic and related rocks.* John Wiley and Sons, Inc., New York. 474.

Chapter 6

Stratigraphic classification

In recent years there has been a major effort to reach international agreement on the principles of stratigraphical classification. Reports have been published by the International Subcommission on Stratigraphic Classification of the International Union of Geological Sciences (Hedberg, 1976) and the Stratigraphy Committee of the Geological Society of London (Holland, *et al.,* 1978).

Stratigraphy is the study of rock successions, whether bedded or unbedded, sedimentary, metamorphic or igneous. Stratigraphers describe rock successions and their relationships, and interpret them to determine sequences of environments and events in the geological history of the earth. Determining the age and the timing of these events is of paramount importance.

As far as classification is concerned many different methods or criteria may be used to divide and group rock strata into distinctive units, for example, lithology, fossil content, radiometric dating, seismic character, magnetic polarity, etc. The Subcommission recognised the following three categories as the best known and the most widely used (Hedberg, 1976, p.):

1. *Lithostratigraphy* — that element of stratigraphy which is concerned with the organisation of strata into units based on their lithologic character.

2. *Biostratigraphy* – that element of stratigraphy which is concerned with the organisation of strata into units based on fossil content.
3. *Chronostratigraphy* – that element of stratigraphy, which is concerned with the organisation of strata into units based on their age relations. The essential point of chronostratigraphy is that the boundaries of the unit should be isochronous.

6.1 LITHOSTRATIGRAPHY

Lithostratigraphy, that is, rock stratigraphy: the description of local rock successions in terms of petrological, mineralogical or general palaeontological characters. Lithological units may be marked by sharp boundaries or by selected points in a gradational sequence.

Hierarchical classification of lithostratigraphical units.

Supergroup – two or more adjacent groups.

Group – two or more adjacent formations.

Formation – the basic subdivision which can be mapped at the surface or traced subsurface. A formation should show some internal lithological homogeneity or possess, lithological features which distinguish it from adjacent strata.

Member – part of a formation, distinguished by its lithological distinctiveness.

Bed – the smallest lithostratigraphical unit, a named distinctive layer. (In older literature the term Bed is equivalent to Formation).

Lithostratigraphical units should be defined at one or more type section, which are precisely located geographically. The units should be described lithologically. Names are generally taken from a geographical feature at or near the type section, for example, London Clay Formation. In Britain the Institute of Geological Sciences keeps an updated master file of British stratigraphical names in order that newly created names will not duplicate those already used.

6.2 BIOSTRATIGRAPHY

Biostratigraphy is the application of fossils to stratigraphical subdivision and correlation. The basic biostratigraphical unit is a biozone which takes its name from one or more index fossils, for example, *Globorotalia truncatulinoides* Biozone. Several types of biozone are recognised (see Hedberg, 1976, or Holland, *et al.*, 1978). Some authors treat biozones as chronozones (for example, ammonite zones in the Jurassic, planktonic foraminiferal zones in the Cenozoic) but the concept of the two types of zone is different. Biozones are limited on application to the geographical distribution of the index species. Thus they are never of global application.

6.3 CHRONOSTRATIGRAPHY

Chronostratigraphy deals with the age of strata and their time relations. Chronostratigraphical classification of rock strata into isochronous units is ideally of global application, unlike lithostratigraphical and biostratigraphical units which are never global in extent. The relative time scale, initiated more than a century ago, is still being refined. The hierarchical classification of units is shown below:

Chronostratigraphical Unit	Geochronological
† Eonothem	Eon
† Erathem	Era
System	Period
Series	Epoch
Stage	Age
Chronozone	† Chron

(† = rarely used in practice).

In general the smallest unit used is the stage. To avoid overlap in time between chronostratigraphical units, the Subcommission recommends that each unit should be defined at its base by a 'boundary stratotype', that is, a lithological succession at a specific geographical locality. The top of a chronostratigraphical unit is defined by the base of the unit above.

Ideally, boundary-stratotypes of stages should be within marine sequences of continuous deposition and should be associated with biozone boundaries or other distinct isochronous horizons. Unfortunately, many stages fail to meet these requirements. Indeed, the ideal of a global scheme of stages is far from being realised. At present each continent has its own local stages, often with separate schemes for marine and non-marine successions. Chronostratigraphical correlation between these local stages is still a matter for research and debate. International attention is at present focused on the definition of boundary stratotypes for systems. The boundary of a system must by definition coincide with the appropriate division of lowest rank.

The major units of chronostratigraphy are shown in Fig. 6.1. The Tertiary is commonly divided into Palaeogene and Neogene with the status of Periods or sub-Periods. The Series in the Cenozoic are given names. For the Mesozoic and Palaeozoic the Series are represented by by the divisions lower, middle and upper and the Epochs early, middle and late. Details of local stages for Europe, North America, New Zealand and S.E. Australia are given by Van Eysinga (1975).

Comment
Although the Geological Society of London has endorsed the proposals of the Subcommission by publishing its own *Guide to stratigraphical procedure* (Holland *et al.,* 1978)

Eonothems & eons	Erathems & eras	Systems & Periods		Series & Epoch
Phanerozoic	Cenozoic	Quaternary		Holocene Pleistocene
		Tertiary	Neogene	Pliocene Miocene
			Palaeogene	Oligocene Eocene Palaeocene
	Mesozoic	Cretaceous Jurassic Triassic		
	Palaeozoic	Permian Carboniferous Devonian Silurian Ordovician Cambrian		
Precambrian or Archaeozoic or Cryptozoic				

Fig. 6.1 – Major chronostratigraphical units. Left of diagram after Hedberg, 1976; right after Van Eysinga, 1975.

it should be appreciated that the traditional approach of British geologists to stratigraphy differs in certain respects. For instance biozones have been held to be time related and their boundaries to be isochronous (*see* Donovan, 1966). Also, only recently have the terms Group, Formation and Member come into use. The dual use of old and new nomenclature will continue for many years.

6.4 REFERENCES

Donovan, D. T. 1966. *Stratigraphy. An introduction to principles.* Murby, London.

Hedburg, H. D. (Editor) 1976. *International Stratigraphic Guide. A guide to stratigraphic classification, terminology and procedure.* John Wiley and Sons. New York. pp. 200.

Holland, C. H. *et al.*, 1978. A guide to stratigraphical procedure. *Geol. Soc. Lond. Spec. Rep.* No.10, pp. 18.

Van Eysinga, F. W. B. 1975. *Geological Time Scale.* 3rd edition. Elsevier.

Chapter 7

Engineering Geology

Civil Engineers expect geological information to be presented in a fashion which is relevant to their own needs. Standard geological classification schemes are often not suitable because they contain too much detail not necessary for the engineer, and often they omit information which is important to the engineer. Consequently, separate engineering geological classification schemes have been erected.

In a recent report by the Geological Society Engineering Group Working Party on *The description of rock masses for engineering purposes* (Chaplow, *et al.*, 1977) the classification of rock material proposed by Dearman, was recommended (see Figs 7.1. and 7.2.). This was set up to provide an identification without a detailed petrographic analysis. The terms used do not all have the more precise meanings of the strictly geological classifications.

Traditionally, engineers have distinguished between rock and soil, the latter being characterised by disintegrating by gentle mechanical means, for example, agitation in water. Engineering soils are tested according to British Standards Institution BS 1377 (1975). From the engineering geological point of view the division into rock and soil is regarded as subjective and Chaplow, *et al.*, (1977) recommend that there should be a unified approach to the description and classification of all rock materials.

GENETIC GROUP	DETRITAL SEDIMENTARY			CHEMICAL/ ORGANIC	PYROCLASTIC
Usual Structure	Bedded				Bedded
Composition	Grains of rock, quartz, feldspar and other minerals		At least 50% carbonate grains		
Grain size mm					
60 — Very coarse	Conglomerate: rounded rock fragments	Rudaceous	Calcirudite		At least 50% of grains are fine-grained volcanic rock
Coarse	Breccia: angular rock fragments	Rudaceous	Calcirudite		Agglomerate: grains rounded; Volcanic breccia: grains angular
2 — Medium	Sandstone: grains are mainly mineral fragments; Quartz sandstone: 95% quartz; Arkose: 75% quartz, up to 25% feldspar; Argillaceous sandstone: 75% quartz + fine detrital material	Arenaceous	Calcarenite	Saline rocks Halite, anhydrite, gypsum	Tuff
0.06 — Fine	Mudstone; Shale: fissile mudstone; Siltstone: 50% fine-grained particles; Claystone: 50% very fine-grained particles; Calcareous mudstone	Argillaceous or Lutaceous	Calcisiltite	Chert, Flint; Coal	Fine-grained tuff
0.002 — Very fine	(as above)	Argillaceous or Lutaceous	Calcilutite		Very fine-grained tuff

Limestones – undifferentiated (spans carbonate column)

Fig. 7.1 – Classification of sedimentary and pyroclastic rocks for engineering geological purposes (from Chaplow, *et al.*, 1977, modified from Dearman).

GENETIC GROUP	IGNEOUS					METAMORPHIC	
Usual structure	Massive					Foliated	Massive
Composition	Light-coloured minerals are quartz, feldspar, mica, and feldspar-like minerals				Dark minerals	Quartz, feldspar, mica, acicular dark minerals	
Grain Size mm	Acid rocks	Intermediate rocks	Basic rocks		Ultrabasic		
Very coarse (60)	Pegmatite				Pyroxenite and Peridotite	Migmatite	Hornfels / Marble
Coarse	Granite	Diorite	Gabbro			Gneiss: alternate layers of light and dark minerals	Granulite
Medium (2)	Micro-granite	Micro-diorite	Dolerite		Serpentinite	Schist	Quartzite / Amphibolite
						Phyllite	
Fine (0.06)	Rhyolite	Andesite	Basalt			Slate	
Very Fine (0.002)						Mylonite	
Glassy	Obsidian and pitchstone		Tachylite				

Fig. 7.2 – Classification of igneous and metamorphic rocks for engineering geological purposes (from Chaplow, *et al.*, 1977, modified from Dearman).

7.1 ROCK MATERIALS

Description of rock material should include:

Group 1: descriptive indices

    Rock type (Figs 7.1. and 7.2.)
    Colour
    Grain size (Fig. 7.3.)
    Texture and fabric, for example, glassy, granular, crystalline.
    Weathering and alteration (Fig. 7.4.)
    Strength (Fig. 7.5.)

Group 2: indices determined by classification tests requiring little or no sample preparation

    Hardness
    Durability
    Porosity
    Density
    Strength
    Sonic velocity

PARTICLE SIZE	DESCRIPTIVE TERM	EQUIVALENT SOIL GRADE
> 60 mm	Very coarse – grained	Boulders, cobbles
2 – 60 mm	Coarse – grained	Gravel
60 μm – 2 mm	Medium – grained	Sand
2 – 60 μm	Fine – grained	Silt
< 2 μm	Very fine – grained	Clay

Fig. 7.3 – Classification of grain size for engineering purposes (after Chaplow, *et al.*, 1977).

NOT DISCOLOURED	DISCOLOURED					
	On major discontinuity surfaces only	On discontinuity surfaces and rock				
			Decomposition and/or disintegration			
			<50%	>50%	100%	
					Mass structure intact	Mass structure and fabric destroyed
Fresh	Fairly weathered	Slightly weathered	Moderately weathered	Highly weathered	Completely weathered	Residual soil
IA	IB	II	III	IV	V	VI

Fig. 7.4 – Classification of weathering and alteration (adapted from Chaplow, *et al.*, 1977).

Descriptive term	Unconfined compressive strength MN/m^2 (MPa)	Field estimate of hardness
Very strong	>100	Very hard rock, more than one hammer blow required to break specimen
Strong	50-100	Hard rock; hand held specimen can be broken with a single hammer blow
Moderately strong	12.5-50	Soft rock; 5 mm indentations with sharp end of pick
Moderately weak	5.0-12.5	Too hard to cut by hand
Weak	1.25-5.0	Very soft rock; material crumbles under firm blow with the sharp end of geological pick
Very weak rock or hard soil	0.60-1.25	Brittle or tough, may be broken in the hand with difficulty
Very stiff	0.30-0.60†	Soil can be indented by the finger nail
Stiff	0.15-0.30	Soil cannot be moulded in fingers
Firm	0.08-0.15	Soil can be moulded only by strong pressure of fingers
Soft	0.04-0.08	Soil easily moulded with fingers
Very soft	<0.04	Soil exudes between fingers when squeezed in hand

Fig. 7.5 – Classification of rock material strength (from Chaplow, *et al.*, 1977).   † Values for soils are double the unconfined shear strength.

Group 3: indices determined by complex testing and/or requiring extensive sample preparation

Young's modulus of elasticity
Poisson's ratio
Primary permeability

Only the descriptive indices of group 1 fall within the scope of this book.

7.2 ROCK MASSES

Description of rock masses is based on the structure and distribution of different rock types and the associated weathering profile.

Group 1: descriptive indices

Discontinuities. "A discontinuity is considered to be a plane of weakness within the rock across which the rock material is structurally discontinuous and has zero or low tensile strength, or tensile strength lower than the rock material under the stress levels generally applicable in engineering. Thus a discontinuity is not necessarily a plane of separation." (Chaplow, *et al.* 1977). Examples are faults, joints, fissures, veins, tension cracks, cleavage, schistocity, shear planes, foliations, bedding planes and lamination.

The number of sets, location, orientation (dip and direction of dip), spacing, (Fig. 7.6.) aperture (Fig. 7.6.), nature of the infill and nature of the surfaces (waviness, roughness and condition of the walls) should all be recorded. Three-dimensional discontinuity patterns are described as in Fig. 7.6.

Weathering and alteration (Fig. 7.4.)

Group 2: indices determined by simple classification tests

Secondary permeabilty
Seismic velocity
Shear strength

Rock Masses

DESCRIPTIVE TERM	SPACING	APERTURE (DISCONTINUITIES) THICKNESS (VEINS, FAULTS)	BLOCK SIZE	BLOCK DESCRIPTIVE TERMS
Extremely wide	>2m		>8 m^3	Very large
Very wide	600mm – 2m		0.2 – 8 m^3	Large
Wide	200 – 600 mm	>200 mm	0.008 – 0.2 m^3	Medium
Moderately wide	60 – 200 mm	60 – 200 mm	0.0002 – 0.008m^3	Small
Moderately narrow	20 – 60 mm	20 – 60 mm	<0.0002 m^3	Very small
Narrow	6 – 20 mm	6 – 20 mm		
Very narrow	<6 mm	2 – 6 mm		
Extremely narrow		>0 – 2 mm		
Tight		zero		

Fig. 7.6. – Descriptive terms for discontinuity surfaces (from Chaplow, *et al*, 1977). The mean values should be determined by counting the number and width of discontinuities cutting a traverse of known length. Blocks are formed by the dimensional discontinuity spacings.

Group 3: indices determined by complex testing
    Modulus of deformability
    Secondary permeability
    Seismic velocity
    Shear strength

Only the descriptive indices of group 1 fall within the scope of this book.

7.3. CORE LOGGING

Recommendations for the logging of cores for engineering purposed have been published by Knill, *et al.* (1970) and Price, *et al.* (1977).

7.4. AGGREGATES

The gravel, crushed rock, and sand used in the manufacture of concrete are *aggregates*. Eleven Trade Groups are recognised to avoid detailed petrological identification (Blyth and de Freitas, 1976).

Trade Group	Rock types
1. Artificial	-
2. Basalt	Basalt, andesite, dolerite, hornblende-schist
3. Flint	Flint, chert
4. Gabbro	Gabbro, diorite, norite, peridotite, picrite, serpentinite
5. Granite	Granite, gneiss, granodiorite, pegmatite, quartz-diorite, syenite
6. Gritstone	Grit, agglomerate, conglomerate, breccia, arkose, sandstone, tuff, pumice, greywacke.
7. Hornfels	All contact altered rocks except marble.

8. Limestone	Limestone, dolomite, marble.
9. Porphyry	Porphyry, quartz porphyry, aplite, felsite, granophyre, microgranite, rhyolite, trachyte
10. Quartzite	Orthoquartzite, metaquartzite, ganister.
11. Schist	Schist, phyllite, slate.

Unconsolidated sediments are classed as gravels or sand depending on the size of the grains:

for concrete, sand is <4.76 mm diameter (BS 882, 1965)

for tarmacadam and bitumen macadam, sand is <3.17 mm diameter (BS 1241, 1959 and BS 2040, 1953).

The Mineral Resources Division of the Institute of Geological Sciences take the boundary at 4.00 mm (Archer, 1972).

7.5 REFERENCES

Archer, A. A. 1972 *Sand and gravel as aggregate.* Mineral Dossier No. 4, 29 pp. H. M. S. O.

Blyth, F. G. H. and De Freitas, M. H. 1976 *A geology, for engineers.* Arnold, London. 587 pp.

British Standards Institution, 1975 BS 1377. *Methods of test for soils for civil engineering purposes.* 143 pp.

Chaplow, R., Edmond, J. M., McCann, D, and Rawling G. E. 1977 The description of rock masses for engineering purposes. Report by the Geological Society Engineering Group Working Party. *Q. Jl. Engng Geol.* **10**, 355-388.

Knill, J. L., Cratchley, C. R., Early, K. R., Gallois, R.W., Humphreys, J.D., Newberry, J., Price, D. G. and Thurrell, R. G. 1970 The logging of rock cores for engineering purposes. *Q. Jl. Engng Geol.* **3**, 1-24.

Price, D. G., Gallois, R., Gordon, D. L., Knill, J. L., Lovelock, P. E. R., Newberry, J., Power, T. O. and Rankilor, P. R. 1977 The logging of rock cores for engineering purposes. *Q. Jl. Engng Geol.* **10**, 45-52.

Index

A
achnelith, 28
adamellite, 61
aeolianite, 49
age, 90 - 92
agglomerate, 27, 95, 102
aggregate, 102, 103
agmatic structure, 85
agmatite, 85
alkali-feldspar charnockite, 65
alkali-feldspar granite, 59 - 61, 65
alkali-feldspar rhyolite, 59, 68
alkali-feldspar syenite, 59 - 61, 65
alkali-feldspar trachyte, 59
alkali granite, 61
alkali igneous rock series, 73
alkali rhyolite, 69
alkali trachyte, 69
alkali rocks, 62
allochem, 37, 40
alloclastic, 27, 32
    volcanic breccia, 32
alnoite, 67
alteration, 98
alvikite, 68
amphibolite, 78, 96
analcimite, 72
anatexis, 86
andesite, 59, 69 - 71, 73, 96, 102
anhydrite, 46, 95
anorthosite, 57 - 62
anthracite, 35
aphyric, 66
aplite, 66, 103

Archaeozoic, 92
arenaceous, 15, 95
arenite, 15, 16, 19, 21
    arkosic, 21
    feldspathic, 21
    lithic, 20, 21
    quartz, 21
    volcanic, 27
argillaceous, 15, 95
arkose, 20, 21, 95, 102
ash, 28, 29
asphaltic earth, 34
asphaltite, 34
augen gneiss, 80, 85
augen structure, 85
autoclastic, 27, 31
    volcanic breccia, 27, 31

B
bafflestone, 41, 43
ball clay, 23
basalt, 59, 69 - 71, 96, 102
    alkali, 71
    analcime, 72
    calc-alkaline, 71, 73
    glass, 72
    leucite, 72
    leuco, 72
    nepheline, 72
    tholeiitic, 71
basanite, 59, 69, 71, 73
bauxite, 33
bed, 89
beforsite, 68

Index

bentonite, 23, 27, 30
bindstone, 43
biolithite, 39, 40, 41
biomicrite, 40
biopelmicrite, 40
biopelsparite, 40
biosparite, 40
biostratigraphy, 89, 90
biozone, 90
black shale, 24, 34
blastomylonite, 79
block, 28
boghead cannel, 34
    shale, 34
bomb, 28
boudins, 85
boulder, 15, 17, 97
boulder clay, 18
boundary stratotype, 91
boundstone, 41, 42
breccia, 14, 15, 17, 27, 31, 32, 85, 95, 102
    alloclastic, 27, 32
    autoclastic, 27
    crush, 79
    explosion, 27, 31, 32
    flow, 27, 32
    friction, 27, 31
    intrusion, 27, 32
    intrusive, 27
    pyroclastic, 27, 29
    tectonic, 79
    tuffisite, 27, 32
    tuffisitic, 32
    volcanic, 27
bronzitite, 65
buchite, 78

C

calc-alkali igneous rock series, 73
calc-alkaline basalts, 71, 73
calcarenaceous sand, 22
calcarenite, 44, 95
calcilith, 37
calcilutite, 44, 95
calcirudite, 44, 95
calcisiltite, 44, 95
calcrete, 46
caliche, 46
Cambrian, 92
camptonite, 67
cannel, 34
    coal, 34
    shale, 34
carbonaceous deposits, 33, 34
carbonate, 35 - 45
    rocks, 35 - 45
    sediments, 35 - 45
carbonatite, 66, 68
Carboniferous, 92
carnallite, 46
cataclasite, 78, 79
cataclastic metamorphism, 77
Cenozoic, 91, 92
chalk, 47
charnockite, 65
chemical precipitates, 14, 47, 95
chert, 47, 86, 95, 102
    diatomaceous, 45
    radiolarian, 46
    replacement, 47
china clay, 23
chloralgal, 35
chlorozoan, 35
chron, 90 - 92
chronostratigraphy, 89 - 91
chronozone, 90 - 92
clay, 15, 23, 24, 47, 97
    ball, 23
    china, 23
    kaolinitic, 23
    montmorillonite, 23
    pipe, 23
claystone, 23, 24, 47, 95
clinopyroxenite, 64, 65
coal, 34, 95
    bituminous, 35
    bone, 34
    brown, 35
    rank, 35
    series, 34
cobble, 15, 17, 97

Index

conglomerate, 14, 15 - 18, 27, 79, 95, 102
  crush, 74
  extraformational, 18
  intraformational, 14
  oligomict, 18
  orthoconglomerate, 18
  orthoquartzitic, 18
  paraconglomerate, 18
  petromict, 18
  sharpstone, 17
  tectonic, 79
  volcanic, 27
conglomeratic argillite, 18
conglomeratic mudstone, 18
contact metamorphism, 77
contourite, 49
coralgal, 36
Cretaceous, 92
Cryptozoic, 92
crystal tuff, 29

D
dacite, 59, 69, 70, 73
deep-sea sediments, 47, 48
Devonian, 92
diabase, 66, 73
diallagite, 65
diamictite, 17, 18
diamicton, 18
diatexis, 86
diatexite, 86
diatomaceous, 45, 48
  chert, 45, 48
  ooze, 46, 48
diatomite, 45, 46, 48
dictyonitic structure, 85
dilation structure, 85
diopsidite, 65
diorite, 59 - 61, 65, 73, 96, 102
dioritoid, 57
discontinuity, 100, 101
dismicrite, 39, 40
dolerite, 66, 96, 102
dolomite, 44, 45, 103
dolostone, 44, 45
dunite, 64, 86

E
eclogite, 80
enderbite, 65
engineering geology, 94 - 104
enstatitite, 65
Eocene, 92
eolianite, 49
eon, 90 - 92
eonothem, 90 - 92
epoch, 90 - 92
era, 90 - 92
erathem, 90 - 92
essexite, 62
evaporite, 46
explosion-breccia, 27, 31, 32

F
fanglomerate, 17, 18
fels, 78, 81, 82
felsite, 103
ferocarbonatite, 68
fireclay, 22, 23
fleck structure, 85
flint, 47, 95, 102
floatstone, 43
flow-breccia, 27, 32
foid
  diorite, 59, 62
  gabbro, 59, 62
  monzodiorite, 59, 62
  monzogabbro, 59, 62
  monzosyenite, 59, 62
  syenite, 59, 62
foidite, 59, 69 - 71
foidolite, 57, 59, 60
folded structure, 85
foramol, 35
formation, 89
framestone, 41, 43
friction-breccia, 27, 31
fuller's earth, 23

G
gabbro, 59 - 63, 73, 86, 96, 102
gabbroid, 57, 58, 63
ganister, 103
glass, 29, 72, 79

Index

glassy rocks, 72
gneiss, 78, 80 - 82, 96, 102
grainstone, 42, 43
granite, 59 - 61, 65, 73, 96, 102
granitoid, 57, 59, 60
granodiorite, 59 - 61, 65, 73, 102
granophyre, 66, 103
granule, 15, 17
granulite, 80, 96
grapestone, 36 - 39
gravel, 14, 15, 17, 97, 103
greensand, 22
greenstone, 66, 73
greywacke, 19, 21, 102
grit, 102
gritstone, 102
group, 89
gypsum, 46, 95

H
halite, 46, 95
hartschiefer, 79
harzburgite, 64, 86
hawaiite, 71, 73
Holocene, 92
hornblendite, 58, 64 - 65
hornfels, 78, 96, 102
hyaloclastic, 27, 31
hyaloclastite, 27, 31
hyalomylonite, 79
hyalo-rhyolite, 72
hyalotuff, 27, 31
hydroclastic, 27, 31
hydroclastite, 27
hypabyssal rocks, 55, 65 - 68
hypersthenite, 65

I
igneous rocks, 53 - 75, 96
igneous rock series, 72, 73
ignimbrite, 30, 31
intraclast, 37 - 40
intramicrite, 40
intrasparite, 40
intrusion-breccia, 27, 32
ironstone, 46

J
jotunite, 65
Jurassic, 92

K
kainite, 46
kaolinitic clay, 23
keratophyre, 74
kersantite, 67
kieserite, 46
kimberlite, 87
komatiite, 65

L
lahar paraconglomerate, 27, 33
laminite, 49
lamprophyre, 66, 67
lapilli, 28, 29
lapillistone, 27, 29
lapilli-tuff, 29
larvikite, 61
laterite, 13, 33
latite, 59, 69, 70
leucitite, 72
leuco-basalt, 71
leucosome, 84
lherzolite, 64, 86
lime mud, 36
limestone, 35 - 45, 47, 95, 103
    allochthonous, 43
    autochthonous, 41, 43
    bioherm, 40
    bituminous, 24
    dolomitic, 44, 45
    Dunham classification, 39 - 41
    Embry & Klovan classification 41 - 44
    Folk classification, 37 - 39
    hybrid, 44
    magnesian, 45
    microcrystalline, 40
    sparry, 40
liparite, 68
lithic, 20, 21, 29
    fragments, 20, 21
    tuff, 29
lithostratigraphy, 88 - 90

Index

loess, 22
loessite, 49
lutaceous, 15, 95
lutite, 15, 16, 22, 23, 24
    hybrid, 24

M
magnesian limestone, 45
malingite, 59
mangerite, 65
marble, 78, 81, 82, 96, 103
marl, 24, 44, 47
marlstone, 24, 44, 47
mela-andesite, 71
melanosome, 84
melilitite, 72
member, 89
Mesozoic, 92
meta-, 73, 77
metamorphic rocks, 76 - 83, 96
metamorphism, 77
metaquartzite, 78, 103
metatexis, 86
metatexite, 86
micrite, 24, 37 - 40
microdiorite, 96
microgranite, 66, 96, 103
migmatite, 84 - 86, 96
minette, 67
Miocene, 92
mixtite, 17
monchiquite, 67
montmorillonite clay, 23
monzodiorite, 59, 60, 65, 73
monzogabbro, 59, 60, 73
monzogranite, 61
monzonite, 59 - 61, 65, 73
monzonorite, 65
mud, 23
    lime, 36
mudrock, 22, 23
mudstone, 15, 22, 23, 42, 43, 95
    tuffaceous, 24, 27
    volcanic, 27
mugearite, 71, 73
mylonite, 78, 79, 96

N
natrocarbonatite, 68
nebulitic structure, 86
Neogene, 91, 92
neosome, 84 - 86
nephelinite, 72
norite, 61 - 63, 65, 102

O
obsidian, 72, 96
ocean floor metamorphism, 77
oil shale, 24, 34
Oligocene, 92
oligomict conglomerate, 18
olistolith, 18
olistostrome, 18
olivinite, 65
ooid, 38 - 40
oolith, 38
oolite, 36
oomicrite, 40
oosparite, 40
ooze, 46 - 48
    diatom, 46, 48
    foraminiferal, 48
    Globigerina, 48
    nanno, 48
    pteropod, 48
    radiolarian, 46, 48
    siliceous, 48
opdalite, 65
ophiolite, 86
ophthalmic structure, 85
Ordovician, 92
organic deposits, 13, 33 - 46, 95
orogenic metamorphism, 77
orthoconglomerate, 18
orthogneiss, 80
orthopyroxenite, 64, 65
orthoquartzite, 21, 103
orthoquartzitic conglomerate, 18

P
packstone, 42, 43
Palaeocene, 92
Palaeogene, 91, 92
Palaeozoic, 92

palagonite, 72
palagonite tuff, 27, 31
Paleocene *see* Palaeocene, 92
Paleogene *see* Palaeogene, 91, 92
paleosome, 84 - 86
Paleozoic *see* Palaeozoic, 92
pantellerite, 68
paraconglomerate, 18
    lahar, 27, 33
paragneiss, 80
peat, 33 - 35
pebble, 15, 17
pegmatite, 66, 96, 102
pelite, 15
pelitic, 15
pellet, 36 - 40
pelmicrite, 40
pelsparite, 40
pépérite, 27, 33
peridotite, 58, 64 - 65, 86, 87, 96, 102
period, 90 - 92
Permain, 92
petromict conglomerate, 18
Phanerozoic, 92
phenocrysts, 29, 66
phi (∅) scale, 15
phlebitic structure, 85
phonolite, 59, 69 - 71, 73
phonolitic tephrite, 59, 69
phosphorite, 46
phreatomagmatic, 27, 31
phyllite, 77, 81, 96, 103
picrite, 66, 102
pillow lava, 31, 73, 74, 86
pipe clay, 23
pitchstone, 72, 96
plagio-granite, 61
plagisyenite, 62
Pleistocene, 92
Pliocene, 92
plutonic rocks, 55 - 65
polyhalite, 46
polymict conglomerate, 18
polzenite, 67
porcellanite, 48
porphyritic, 66

porphyrite, 66
porphyroclast, 78
porphyry, 103
Precambrian, 92
protocataclasite, 79
protomylonite, 79
psammite, 15
psammitic, 15
paeudobreccia, 45
pseudotachylite, 79
ptygmatic structure, 85
pulaskite, 61
pumice, 28, 102
pyroclastic, 13, 27 - 31, 95
    agglomerate, 27
    breccia, 27
    fall deposits, 30
    flow deposits, 30
    ground surge deposits, 30
    rocks, 28 - 31
pyroxenite, 58, 64, 65, 86, 96

Q
Quaternary, 92
quartzite, 78, 81, 82, 96, 103
quartzolite, 59, 60
quartzwacke, 21

R
radiolarian, 46, 48
    chert, 46, 48
    ooze, 46, 48
radiolarite, 46, 48
raft structure, 85
regional metamorphism, 77
residual deposits, 13, 83
rhyodacite, 68
rhyolite, 30, 59, 68 - 70, 73, 96, 103
rhythmite, 49
rodingite, 87
rubble, 17
rudaceous, 15, 95
rudite, 14 - 17
rudstone, 43

Index

S

sand, 15, 97, 103
    calcarenaceous, 22
    glauconitic, 22
    skeletal, 35
sandstone, 15, 16, 19, 95, 102
    feldspathic, 19
    hybrid, 20
    maturity, 19, 20
    muddy, 19
    tuffaceous, 22, 27
    volcanic, 27
sannaite, 67
sapropel, 24, 33, 34
schist, 78, 81, 82, 96, 102, 103
schlieren structure, 85
schollen structure, 85
scoria, 28
scree, 17
seatearth, 22, 23
sediment, 13 - 52
    carbonate, 35, 36
    chemical, 46 - 47
    deep-sea, 47, 48
    organic, 33 - 46
    residual, 33
    terrigenous, 13 - 24
    volcanic, 25 - 33
sedimentary rock, 13 - 51, 95
series, 90 - 92
serpentinite, 78, 86, 87, 96, 102
shale, 16, 23, 34, 95
    black, 24, 34
    boghead cannel, 34
    cannel, 34
    carbonaceous, 24
    oil, 34
    organic, 24
sharpstone conglomerate, 17
shonkinite, 59
silcrete, 47
siliceous, 45, 46, 48
    ooze, 48
    rocks, 45, 46
    sediments, 45, 46
silt, 15, 22, 97
siltstone, 15, 22, 95
    tuffaceous, 27
    volcanic, 27
Silurian, 92
skarn, 78
slate, 77, 96, 103
soil, 94, 98, 99
sövite, 66
spar, 37 - 40
spessatite, 67
spilite, 73
stage, 90 - 92
stictolithic structure, 85
stratigraphic classification, 88 - 93
stratotype, 91
stromatic structure, 85
subarkose, 21
sublitharenite, 21
supergroup, 89
surreitic structure, 85
syenite, 59 - 61, 65, 73, 102
syenitoid, 57
sylvite, 46
system, 90 - 92

T

tachylite, 72, 96
tephra, 25, 28
tephrite, 59, 69 - 71
tephritic phonolite, 59, 69
terra rosa, 33
terrigenous, 13 - 24, 27, 32, 33
    deposits, 13 - 24
    rocks, 14
    sediments, 14
    volcanic-rich, 27, 32, 33
Tertiary, 91, 92
teschenite, 62, 66, 73
theralite, 62
thermal metamorphism, 77
tholeiitic basalts, 71, 73
tholeiitic igneous rock series, 73
tidalite, 49
till, 17, 18
tillite, 17, 18
tilloid, 17, 18
tonalite, 59 - 61, 65, 73

Index

tonstein, 23
trachyte, 59, 69 - 71, 73, 103
Triassic, 92
troctolite, 63
trondhjemite, 61
tuff, 27, 95, 102
   crystal, 29
   lithic, 29
   palagonite, 21, 31
   sintered, 30
   vitric, 29
   welded, 30
tuffaceous, 22, 27
   breccia, 29
   mudstone, 27
   sandstone, 22, 27
   siltstone, 27
tuffisite, 27, 32
tuffisitic breccia, 32
tuffite, 22
turbidite, 49

U

ultrabasic rocks, 73, 87, 96
ultracataclasite, 79
ultramafic rocks, 56, 63 - 65, 72, 86
ultramafitite, 59, 72
ultramylonite, 79

V

vein structure, 85
vitric tuff, 29
vitrophyre, 72
volcanic, 13, 25, 27, 55, 68, 74, 95
   arenite, 27
   breccia, 27, 95
   conglomerate, 27
   mudstone, 27
   rocks, 55, 68 - 74
   sandstone, 27
   sedimentary deposits, 13, 25
   siltstone, 27
   wacke, 27
volcaniclastic, 25, 26
vosgesite, 67

W

wacke, 19, 21, 27
   arkosic, 21
   feldspathic, 21
   lithic, 21
   volcanic, 27
wackestone, 42, 43
weathering, 98, 100
websterite, 64
wehrlite, 64
Wentworth scale, 15